Copernicus Books
Sparking Curiosity and Explaining the World

Drawing inspiration from their Renaissance namesake, Copernicus books revolve around scientific curiosity and discovery. Authored by experts from around the world, our books strive to break down barriers and make scientific knowledge more accessible to the public, tackling modern concepts and technologies in a nontechnical and engaging way. Copernicus books are always written with the lay reader in mind, offering introductory forays into different fields to show how the world of science is transforming our daily lives. From astronomy to medicine, business to biology, you will find herein an enriching collection of literature that answers your questions and inspires you to ask even more.

Robert T. Stewart

Adventures in Statistics

How We Live in a World of Numbers

 Springer

Robert T. Stewart
Molloy University
Rockville Centre, NY, USA

ISSN 2731-8982   ISSN 2731-8990   (electronic)
Copernicus Books
ISBN 978-3-031-61283-1   ISBN 978-3-031-61284-8   (eBook)
https://doi.org/10.1007/978-3-031-61284-8

This Springer imprint is published by the registered company Springer Nature Switzerland AG
The registered company address is: Gewerbestrasse 11, 6330 Cham, Switzerland

If disposing of this product, please recycle the paper.

Acknowledgments

This book is dedicated to my parents, Beth and Buzz Stewart. Two life-long educators whose love and support made this adventure possible. Thanks.

Contents

1

Average

You Are Not Average

The classic 1990 gangster movie *Goodfellas* ends with Ray Liotta's Henry Hill breaking the fourth wall to complain about life outside the mob: "And that's the hardest part. Today everything is different; there's no action ... have to wait around like everyone else. Can't even get decent food—right after I got here, I ordered some spaghetti with marinara sauce, and I got egg noodles and ketchup. I'm an average nobody ... get to live the rest of my life like a schnook." The nights skirting the Copacabana line with his beautiful date Karen and being set up in the front row of a Henny Youngman show while The Crystals sing "Then He Kissed Me" are gone. Henry Hill is an "average nobody," egg noodles and ketchup.

Average is negative. Nobody wants to come in the middle of the pack, a 0.500 record, a 2.5 GPA. Michael Jordan summed up this feeling in his famous quote, "All I knew is that I never wanted to be average." Even one of the bestselling authors of all time, Anne Frank, questioned why anyone would read the writings of an average girl, "it seems to me that later on neither I nor anyone else will be interested in the musings of a thirteen-year-old schoolgirl."[1] Stop. While fears of egg noodles and ketchup are well grounded, fears of being average are misguided. Nobody is average.

Averages began in the 1700s through use by astronomers. During that time, the crude equipment caused measurement errors. Astronomers tackled this problem using averages; they took many measurements and divided by

---

[1] Frank, Anne. *The Diary of Anne Frank: The Definitive Edition,* p. 24. Anchor Books, 2001.

© The Author(s), under exclusive license to Springer Nature Switzerland AG 2024
R. T. Stewart, *Adventures in Statistics*, Copernicus Books ,
https://doi.org/10.1007/978-3-031-61284-8_1

the number of measurements. This method proved more accurate than any of the faulty measurements alone.[2]

Then, in the early 1800s, a Belgian mathematics teacher named Adolphe Quetelet started using averages for something completely different—studying human behavior. Quetelet computed the average yearly number of births, deaths, suicides, marriages, divorces, and incidences of crime. These were scandalous subjects for his time because people believed that this was the province of God. But Quetelet found that averages could be predicted; he found a pattern; he found the normal distribution. Scholars were thunderstruck. This pattern of normal distributions, centered on these magic averages, initiated debate and controversy. How can "God's will" have such regular statistics?

But Quetelet's most famous work revolved around measurements of the human body.[3] He measured Scottish soldiers including their height, the circumference of their chests, and the length of their arms. Through this exercise, Quetelet developed his philosophy of the "average man" as the ideal, or perfection. He describes the human ideal not to the extremes or highest ends of the distribution (the tallest people), but to the middle or most likely to occur or average in the distribution (the guy who is 5 feet, 9 inches tall). Quetelet's "average man" is not characterized by disapproval like Michael Jordan's quip, but by perfection parallel to Plato's theory of forms. Quetelet's work made him a celebrity around the world, and well known to leaders of the day including Karl Marx, James Garfield, and Abraham Lincoln.

Lincoln followed Quetelet's lead and ordered a massive study of his soldiers. The military began computing averages for everything. Then, they used averages to design weapons, ration food, and create uniform sizes for mass production. Using averages helped secure the North's victory in the Civil War.[4] This success using averages stuck with the military, and in 1926 the Army designed a plane to fit Quetelet's "average man." The army computed averages for all kinds of body characteristics of soldiers, including heights, chest circumference, arm length, crotch height, waist circumference, and neck circumference. Then, the Army designed a cockpit to fit the "average man" such that the distance from the seat to the pedals would fit the average man's body.

[2] Porter, Theodore M. *The rise of statistical thinking, 1820–1900.* Princeton University Press, 2020.

[3] Donnelly, Kevin. *Adolphe Quetelet, social physics and the average men of science, 1796–1874.* Routledge, 2015.

[4] Rose, Todd. *The end of average: How to succeed in a world that values sameness.* Penguin UK, 2016.

But something happened on the way to the victory parade. Crashes. Accidents. Death. The planes were not safe and the after-action reviews (AARs)—designed to determine and fix mistakes—could not explain the failures. That changed when a young Harvard graduate, Lieutenant (Lt.) Gilbert S. Daniels, studied the problem by measuring 4063 pilots on ten size dimensions.

Lt. Daniels measured height, chest circumference, sleeve length, crotch height, torso circumference, hip circumference, neck circumference, waist circumference, thigh circumference, and crotch length.[5] Then, Lt. Daniels computed an average and a range for each dimension. He computed a range such that 25–30% of the pilots would fit. For height, this translated to a range between 173.95 centimeters and 177.95 centimeters tall (greater than 5 feet, 7 inches and less than 5 feet, 11 inches approximately) where 1055 pilots fell according to his measurements. Then, Lt. Daniels chose all the pilots with an average chest circumference which measured between 96.95 centimeters and 100.95 centimeters (roughly 3 feet, 2 inches and 3 feet, 4 inches). Of the 1055 average-height pilots, 302 also had an average chest circumference. Lt. Daniels continued this process as follows:

1. Of the original 4063 pilots, 1055 were of average height.
2. Of these 1055 remaining pilots, 302 had average chest circumference.
3. Of these 302 remaining pilots, 143 had average sleeve length.
4. Of these 143 remaining pilots, 73 had average crotch height.
5. Of these 73 remaining pilots, 28 had average torso length.
6. Of these 28 remaining pilots, 12 had average hip circumference.
7. Of these 12 remaining pilots, 6 had average neck circumference.
8. Of these 6 remaining pilots, 3 had average waist circumference.
9. Of these 3 remaining pilots, 2 had average thigh circumference.
10. Of these 2 remaining pilots, 0 had average crotch length.

Nobody is average. The army built the plane's cockpit to fit Quetelet's ideal "average man" and zero pilots fit the cockpit. Zero pilots fit the cockpit because nobody is average. The one-size-fits-all approach resulted in crashes.

The military hierarchy realized that average cockpits worked for exactly nobody. Making quick decisions with your controls in high pressure situations is what Air Force flying demands. So when the controls don't fit, mistakes happen, death happens. The Air Force informed their suppliers and

---

[5] Daniels, Gilbert S. *The "Average Man."*?. No. TN_WCRD-53-7. Air Force Aerospace Medical Research Lab Wright-Patterson AFB OH, 1952.

insisted on changes. Soon, planes came equipped with adjustable seats and foot pedals. And helmets and flight suits changed designs to fit the dispersion found in the data rather than the average. Crashes decreased. Cars soon followed suit, and now we take adjustable mirrors and seats for granted. Moreover, individuals who used to be immediately eliminated from becoming pilots—including women, but also relatively short or tall men—became pilots such that the Air Force began selecting the best, not average schnooks.

Henry Hill, Michael Jordan, and Anne Frank needn't have worried. They could not have been average because nobody is average. An average is an arbitrary concept that cannot explain any individual. Like snowflakes and diamonds, people are unique. Egg noodles and ketchup are average, but you are not.

Averages Mislead

Larry Summers was never average. At age 16, he started college at the Massachusetts Institute of Technology (MIT). At 28, he became a tenured Harvard professor in economics. Then he served as the Secretary of the Treasury under Bill Clinton before landing a job as the 27th president of Harvard University. Nice resumé. But in January 2005, Larry gave a fateful presentation at an academic conference. He discussed the underrepresentation of females in tenured positions in science and engineering at top universities. Summers downplayed discrimination while offering two hypotheses to explain the low representation. First, he suggested less commitment by women to pursue work over other interests including childcare, and second, he spoke about the "variability of aptitude" which translates into a world that has a lot more really smart men than really smart women.

Larry's comments were not received well. One MIT professor walked out on the speech, saying in the Boston Globe that if she hadn't left, "I would've either blacked out or thrown up."[6] Larry apologized. But national and international news services picked up the story and Larry's resumé took a serious hit. Many believe that the comments cost him his job as president of Harvard and another chance to serve as Secretary of the Treasury under Barack Obama.

We get caught up in differences… How are men different from women? How are Americans different from Mexicans? How are whites different from blacks? How are Jews different from Christians who are different from

---

[6] Hirsh, Michael. *The Comprehensive Case Against Larry Summers*. Atlantic, September 13, 2013.

Muslims? Hans Rosling in his book *Factfulness*, describes this tendency to search for differences as the "gap instinct." He warns of using averages to promote the "gap instinct" when a more representative picture will likely show overlaps rather than gaps, similarities rather than differences.[7]

But gaps and extremes sell. The 4 feet, 8-inch-tall Simone Biles set an American record in the 2016 Olympics by winning four gold medals. The 7 feet, 1-inch-tall Shaquille O'Neal earned his Basketball Hall of Fame induction by winning four NBA championships. They met at the 2017 Super Bowl. Simone posted a picture of them arm-in-arm on Instagram. The post garnered 300,000 likes in 4 hours. The photo is funny and interesting because of the visual representation of the extremes. Still, the majority of people fall in the overlap, not the extremes. We're more alike than different.

A well-known "gap" in average math SAT scores exists between males and females. This anomaly has been used by journalists for a perfunctory story on a slow news day. The average math SAT scores for males and females from 1972 through 2016.[8]

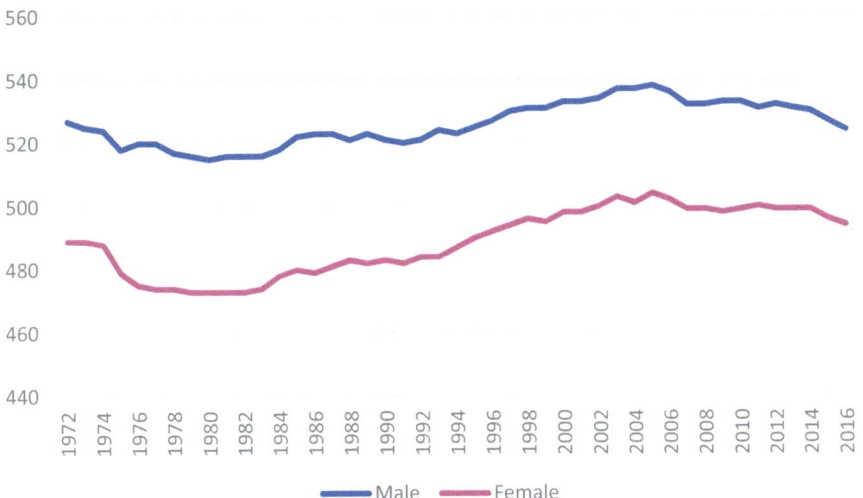

And here is the same picture:

[7] Rosling, H. with O. Rosling and A. Rosling Ronnlund. *Factfulness: Ten reasons we're wrong about the world—and why things are better than you think.* Flatiron, NY 2018.

[8] College Board, SAT Total Group Profile Report.

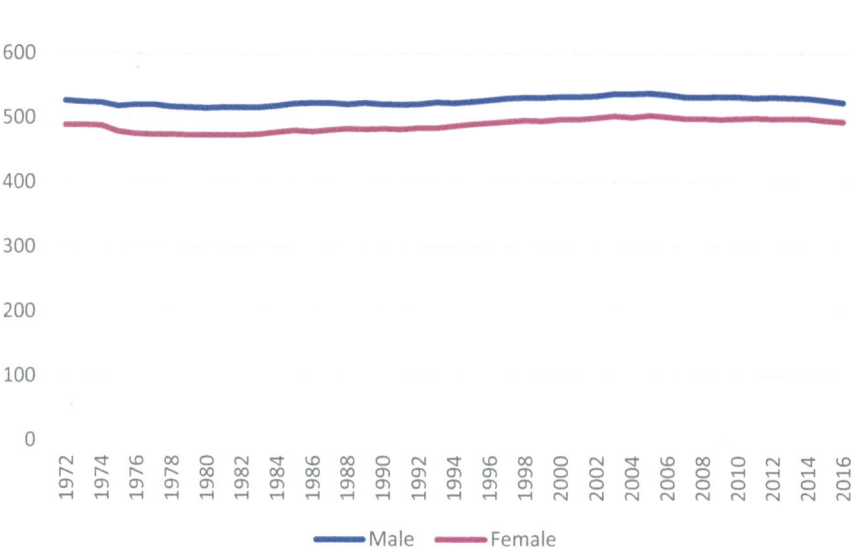

The vertical axis on the top chart starts at 440, while the vertical axis on the bottom chart starts at 0. Otherwise, the charts are identical. You can change the vertical axis to make differences look more or less dramatic. Our minds read the top chart differently from the bottom chart despite both charts using the same data.

Averages never give a full picture. Averages are a summary, a simplification. The male and female math SAT scores broken down into buckets for 2013 data:

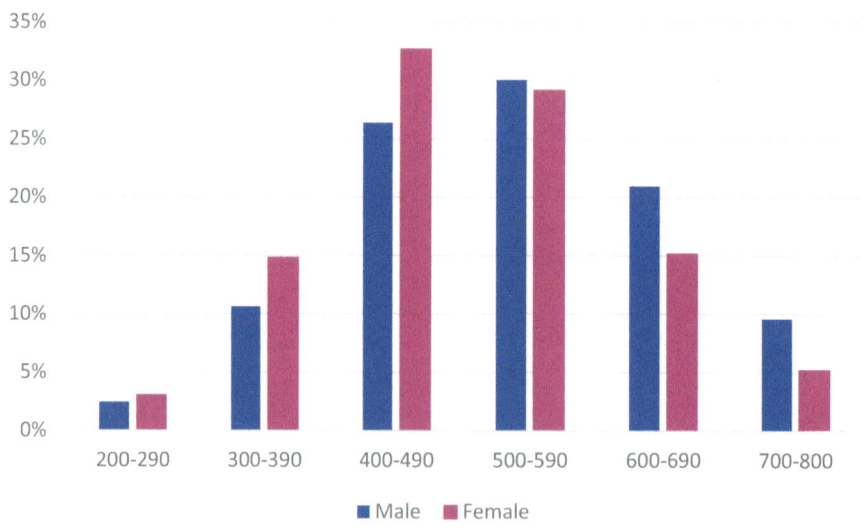

The numbers overlap. Boys and girls have analogous scores. You have to work hard to find a controversial story using this graph.

Still, the graph shows that 10% of the boys scored higher than 700 compared to only 5% of the girls. But more girls take the SAT than boys, and more boys drop out of high school than girls. The population of girls taking the SAT in 2013 was 15% higher than the boys. We're not comparing apples to apples. Or maybe girls are socialized to take math less seriously. Or maybe gender discrimination in mathematics fields has an effect. Who knows? It's tough to make intelligent conclusions with so many factors involved, as Larry Summers learned the hard way. Regardless, the overlap speaks louder than the gaps.

The average American family has 1.9 children. The prodigal statistician has his head in the oven and feet in the freezer but feels fine on average. When Mark Zuckerberg walks onto the rec league basketball team, the average net worth of the players is in the billions. Averages oversimplify. Averages hide a spread or range of different numbers in a single number. Averages make complicated topics seem simple. It's never as easy as "boys are good at math." Be careful, be wary of the gap instinct. Ignore averages dominated by a few observations. And be on the lookout for journalists and politicians selling stories. Averages mislead.

Averages Confound

In 1995, Derek Jeter made the New York Yankee roster to begin his Hall of Fame Career.[9] He scratched out 12 hits in 48 at bats for a batting average of .250 in that year. He followed his meager 1995 debut with an impressive 183 hits in 582 at bats in 1996 for a batting average of .314.

Player	Year	Hits	At Bats	Batting average
Derek Jeter	1995	12	48	.250
	1996	183	582	.314

David Justice was a star for the Atlanta Braves who made consecutive all-star teams in 1993 and 1994. His stats in 1995 and 1996:

Player	Year	Hits	At bats	Batting average
David justice	1995	104	411	.253
	1996	45	140	.321

[9] Ross, Ken. *A Mathematician at the Ballpark: Odds and Probabilities for Baseball Fans.* Penguin, 2007.

David Justice is a better hitter than Derek Jeter and the stats prove it. David has a higher batting average in both 1995 and 1996. But look at the combined years:

Player	Years	Hits	At bats	Batting average
Derek Jeter	1995–1996	195	630	.310
David justice	1995–1996	149	551	.270

Derek Jeter is a better hitter than David Justice and the stats prove it. Derek has a higher batting average from 1995 through 1996. The numbers that we use for definitive decision making are not definitive. When a trend reverses because of how the data is aggregated we get what statisticians call Simpson's paradox. Simpson's paradox happens because a confounding factor is being ignored meaning that something important is lacking from the analysis.

A problem with sample sizes caused the batting average results. Derek Jeter barely played in his inaugural season where he got off to a slow start, which led to few at bats. He was still considered a "rookie" in 1996 when he won the rookie of the year award after an outstanding season. David Justice played a full season in 1995 but missed most of 1996. After a great start in 1996, he dislocated his previously injured right shoulder swinging at a Denny Neagle pitch in May. He had surgery and missed the rest of the season after only 140 at bats.

Who is the better hitter? Single statistics don't tell us much. David Justice had 30 home runs in 1995 and 1996 combined, while Derek Jeter had only 10. You must use statistics like evidence in a jury trial. No one piece of evidence should lead to conviction beyond a reasonable doubt. Smoking guns only exist in the movies. Use multiple statistics combined with multiple hypotheses to determine guilt or innocence. Otherwise, you may find that smoking is good for your health.

"There is nothing more deceptive than an obvious fact," said Sherlock Holmes.[10] From 1972 through 1974 in Newcastle, England, 1314 women took part in a study of heart disease and smoking.[11] Thirty years later a follow-up study found the following average survival rates broken down by smokers and non-smokers:

All ages	Year 1974 total	Year 2004 still alive	Survival rate (%)
Smokers	582	443	76
Non-smokers	732	502	69

---

[10] Doyle, Arthur Conan. *The Adventures of Sherlock Holmes*. Wordsworth Editions, 1992.

[11] Appleton, David R., Joyce M. French, and Mark PJ Vanderpump. "Ignoring a covariate: An example of Simpson's paradox." *The American Statistician* 50, no. 4 (1996): 340–341.

Smokers have a larger survival rate. This is empirical evidence: cold, hard facts. Numbers. This is science. Smoking is good for you. This is an obvious fact.

Unfortunately, something is missing. There is a confounding factor. The above survival rate fails to account for age. And smokers do not live long. The smokers in 1974 consisted of younger women relative to the non-smokers in 1974 leading to the bizarre results. When the smokers and non-smokers are broken down by age, the survival rates flip—Simpson's paradox. Smoking is not good for you. For example, the survival rates for smokers and non-smokers for women aged 55–64 looks as follows:

Ages 55–64	Year 1974 total	Year 2004 still alive	Survival rate (%)
Smokers	115	64	56
Non-smokers	121	81	67

Sherlock Holmes's insight that "there is nothing more dangerous than an obvious fact" should accompany all averages and all statistics. Simpson's paradox is always lurking. Confounding factors that can flip results may be out there just waiting to confuse. Smoking looks healthy… until age enters the picture. The variables that matter the most are often unknown.

Average People Innovate

Innovation is the process of making breakthrough inventions available, affordable, and reliable, according to Matt Ridley in his book *How Innovation Works*.[12] He illustrates his point with a joke. A beaver and a rabbit are looking at the Hoover Dam. The beaver says: "I didn't build it, but it is based on an idea of mine." Invention is the easy part, innovation the hard part. Ridley argues that many of the great innovations throughout history were not done by scientists or the elites. "So, yet again, innovation proves to be gradual and to begin with the unlettered and ordinary people, before the elites take the credit." Average people innovate.

Wilbur Wright was smashed in the face with a hockey stick as a teenager while playing in a neighborhood game.[13] He lost his front teeth. And he lost his self-esteem. His thoughts of attending Yale University disappeared, depression took over, and he spent years largely housebound. His brother Orville Wright dropped out of high school his junior year to start a printing shop.

[12] Ridley, Matt. "How Innovation Works: And Why It Flourishes in Freedom." Harper Collins Publishers, 2020.
[13] McCullough, William. "The Wright Brothers." Simon and Schuster, 2015.

The brothers united in 1892 to open a bicycle shop in Dayton Ohio. The Wright brothers were average.

Samuel Langley's Wikipedia page reads:

Samuel Pierpont Langley was an American astronomer, physicist, inventor of the bolometer, and aviation pioneer. In addition to becoming the third Secretary of the Smithsonian Institution, he was also a professor of astronomy at the University of Pittsburgh.

In 1886 Samuel Langley won both the Rumford Medal by the Royal Society of London, and the Draper Medal of the National Academy of Sciences, and in 1893 he won the Janssen prize of the Paris Academy. Langley built an unpiloted, 25-pound, "heavier-than-air" machine that made two flights of 700 and 1000 meters after a catapult launch on the Potomac River in 1896. This led to Langley receiving a $50,000 government grant from the War Department and $20,000 from the Smithsonian to develop a piloted airplane. Samuel Pierpont Langley was not average.

On December 9th 1903, Langley made his second attempt at manned flight in the Potomac River. The plane collapsed during launch. The press mocked the failure calling it "Langley's folly." A congressman said: "You tell Langley for me…that the only thing he ever made fly was government money." Eight days later, the Wright brothers flew their plane in Kitty Hawk, North Carolina. The estimated cost of their plane was $1000.

The average beat the rich, the famous, the educated, and the exalted. Ripley argues that three features drive innovation: socialization, trial-and-error, and luck. Socialization: a single person rarely innovates; it takes a village. Trial-and-error: Innovation is an evolutionary process; intelligent design does not exist. Luck: serendipity plays a big part in innovation.

Langley insisted on secrecy. No ideas about flying were shared outside his small circle. The Wright brothers corresponded with scientists and flight enthusiasts around the world to share their progress and ideas. Their employee, Charles Taylor, built the plane engine because the Wright brothers knew little about engines. The Wright brothers socialized.

With limited formal education and little faith in equations, the Wright brothers embraced trial-and-error. Langley trusted his mathematics. The Wright brothers built a wind tunnel using a simple wooden box, a square glass window, a fan with a one-horsepower engine, old hacksaw blades, and bicycle spokes. They experimented with lift and drag. They failed. They tried again. They failed again. They argued. They sought help. They failed. They learned. They tried again. They argued. They failed. They tried again. Then, they flew.

Average people innovate. Ridley goes through countless examples in his book including farming, vaccinations, computers, automobiles, lightbulbs, nuclear power, wheeled suitcases, and trains. Again and again, great men don't look so great. Who invented the personal computer? Steve Jobs and Bill Gates played a part. But countless tinkerers and coders and workers made personal computers what they are today. The isolated genius foreseeing the future and creating great inventions does not exist. Top-down planning fails. A lot of average people working together, trying, failing, and trying again leads to innovation, leads to flight. Average people create change.

Average Is Beautiful

Alice Walker won the Pulitzer Prize for Fiction in 1983 for *The Color Purple*. Nettie says in the book, "I think it pisses God off if you walk by the color purple in a field somewhere and don't notice it."[14] Nobody can be sure what pisses God off, but everyone agrees that the color purple in a field is beautiful. We have an innate sense of beauty. So, ignoring purple in fields likely pisses God off, but what is beautiful?

In the 1880s, Sir Francis Galton created the first "composite" image. Through trial-and-error and serendipity, he learned how to project many faces onto a single piece of photographic film. He was hoping to find facial characteristics unique to criminals. He failed. Rather, he found that the composite faces were "better looking" than individual faces. The composites "averaged out" the large noses and pointy chins, leaving perfectly symmetrical faces.[15] Average is perfect. Beauty is average.

Michael Jordan may not have wanted to be average, but he's had no problem dating "average" women. Supermodels are average. Like composite images, choruses and orchestras balance each other out to hit the notes exactly average—creating beautiful music. And in fields around the world God is busy averaging blues and reds into purples; appreciate the beautiful averages the world provides.

---

[14] Walker, Alice. *The Color Purple*, p. 143. Open Road Media, New York, 2011.
[15] Rhodes, Gillian, Alex Sumich, and Graham Byatt. "Are average facial configurations attractive only because of their symmetry?." *Psychological Science* 10, no. 1 (1999): 52–58.

2

Coin Flips

Sacred Coin Flips

In the second episode of the television series *Breaking Bad*, 25-year-old ne'er-do-well Jesse Pinkman barks at Walter White, "And let me tell you something else. We flipped a coin, okay? You and me. You and me! Coin flip is sacred! Your job is waiting for you in that basement, as per the coin!" Jessie and Mr. White are aspiring methamphetamine entrepreneurs with a problem. Negotiations around improving efficiencies in meth distribution collapsed, so Mr. White used phosphine gas to poison two fellow drug distributors, Emilio and Krazy-8. Jesse and Mr. White must dispose of Emilio's lifeless body and kill Krazy-8, who managed to survive the initial poisoning. With these two jobs, they must decide who gets what task. Neither Jesse nor Mr. White is volunteering to kill Krazy-8, as disposing of Emilio's dead body is decidedly easier on the psyche. Their method for deciding: the coin flip.

Let God, or fate, or the vagaries of New Mexico's dry air decide. Mr. White loses the coin flip, but wimps out in the murder, leading to Jesse's wisdom, "Coin flip is sacred! Your job (murdering Krazy-8) is waiting for you in that basement, as per the coin!" Jesse seems alright with breaking a few boundaries like dealing meth and murdering. But violating the sanctity of the coin flip? Now that's a bridge too far.

Jesse's not alone. Man has looked to the Gods and games of chance to make difficult choices for most of history. The Old Testament equivalent to a coin flip is "casting lots." The story of Jonah and the whale begins with sailors on board a ship when a terrible storm hits. "Then, the sailors said to each other, "Come, let us cast lots to find out who is responsible for this calamity." They

R. T. Stewart, *Adventures in Statistics*, Copernicus Books,
https://doi.org/10.1007/978-3-031-61284-8_2

cast lots and the lot fell on Jonah."[1] So clearly, Jonah is guilty of causing the storm. He was thrown overboard and swallowed by a whale. While in the whale, Jonah prayed to God. And God had the whale spit Jonah up after 3 days. He went on to live a pious life. Mr. White wishes he had it so good. Historians argue about what "lots" are exactly, but "astragali" bones or ankle bones of sheep or goats have been found with high frequency at archeological digs across the world. Historians speculate that these ankle bones are the precursor to dice or coin flips. The New Testament also uses the term "casting lots." After Judas betrayed Jesus, the remaining apostles had to choose a replacement. The apostles "cast lots" to choose among the candidates, and thus chose Mathias as an apostle of Christ. The coin flip is truly sacred.

Today, people—excluding Jesse Pinkman—are less enamored of the idea that God's prescience drives the outcomes of random processes like coin flips. The 1988 Permian Panthers high school football team out of west Texas ended district play in a three-way tie for only two playoff spots. The tiebreaker rules stipulated that a three-way coin flip with the odd man out would decide who made the playoffs. HG "Buzz" Bissenger details this coin flip in his book, *Friday Night Lights*, which later became a successful movie and even more successful television show. As depicted by Billy Bob Thornton in the movie, Coach Gary Gaines flipped a head to secure the Panthers' spot in the playoffs. The coin flip was broadcast live on both television and radio because high school football in Texas is crazy, but do you know the rest of the story? Soon after the first ever coin flip to decide who made the playoffs, the powers that be in west Texas high school football changed the rules. A points system was developed such that head-to-head competition and points scored during games would decide who made the playoffs in the event of future ties. Coach Gaines said in the local newspaper, "It's a fairer type of tiebreaker than the luck of the Irish type of thing. It makes more sense. It's more humane if you want to call it that."[2] No more coin flips in Texas high school football.

But the rest of the world thinks football is soccer, and they do not care what games teenagers play in west Texas. Still, the soccer World Cup powers that be also changed their tiebreaker rules to avoid coin flips in 2016. The "fair play" rule was added which adds the number of yellow and red cards administered in the matches as another tiebreaker during group stages, making the coin flip tiebreaker irrelevant. No more coin flips in World Cup soccer too.

---

[1] New Revised Standard Version, Jonah 1.

[2] Hull, Jonathan (November 23, 2013). HS FOOTBALL: "The Coin Toss" proves to be the lasting legacy of the 1988 season, *Midland Reporter Telegram*.

We once chose Christ's apostles with coin flips, and now we refuse to use coin flips for deciding football games and soccer matches. Antagonism towards the coin flip reached its peak during the 2018 Winter Olympics with a tweet from a disgruntled coin flip loser. The US Olympic Team voted to decide who would be the flag bearer during the opening ceremonies and the vote ended in a tie. The coaches decided that a coin flip would provide the fairest decision. Soon after, speed skater Shani Davis tweeted:

> *I am an American and when I won the 1000m in 2010 I became the first American to 2-peat in that event. @TeamUSA dishonorably tossed a coin to decide its 2018 flag bearer. No problem. I can wait until 2022.*

The coin flip—the sacred coin flip used to determine Jesus Christ's apostles—is now subjugated to "dishonorable" status with more "humane" options. What happened?

History of Coin Flips

A gambler in the seventeenth century happened. An eccentric, well-educated writer by the name Chevalier de Mèrè solicited two mathematicians, Blaise Pascal and Pierre de Fermat, to answer a gambling problem.[3] Chevalier de Mèrè was curious about splitting a pot in an unfinished gambling game. This is known as the "problem of points."

Chevalier de Mèrè did not recruit just anyone to solve his problem. Pascal and Fermat are two of the greatest mathematicians who have ever lived. Pascal was a child prodigy who published an original mathematical treatise at 16, then, invented a calculator. The calculator could only add and subtract, but it was the first one, and in 1644. Not to be outdone, Fermat invented number theory, laid the groundwork for modern day calculus, and discovered the fundamental principle of analytic geometry. Pascal and Fermat were brilliant. But in the seventeenth century, brilliance did not translate into any concepts of chance or probability or randomness—these ideas did not exist. God ran the show.

God determined if the coin landed on heads or tails without debate. The concepts of chance and probability and randomness did not exist for most of

---

[3] Devlin, Keith. *The unfinished game: Pascal, Fermat, and the seventeenth-century letter that made the world modern.* Basic Books, 2010.

history.[4] If God wanted you to die, you died. If God wanted you to be King, you were King. And if God wanted you to win at gambling, you won at gambling. God decides was the logic across the world despite diverse cultures, religions, and histories. The future was beyond man's control is what everyone, regardless of their educational background or intellect, believed. The future was God's game.

Compare that to modern day logic. We routinely predict the future: the chance of rain tomorrow is on every nightly news forecast and weather app; the likelihood that the president or senator or town dog catcher gets re-elected is reported on websites and bet on; the likelihood that you will die in the next year or 5 years or 10 years is computed as part of every life insurance policy; and sports teams across the globe are assigned probabilities to win championships by gambling enthusiasts. This 180-degree turn all started with the problem of points. Chevalier de Mèrè changed how we think.

A simple version of the problem of points: Suppose you and I agree that the winner of five coin flips gets $400. So, you get heads and I get tails… the first flip comes up heads and you're up. The next flip comes up tails, and we're even. The next flip comes up heads and you're up two to one. Then… the game stops… somebody has a heart attack, somebody goes into labor, somebody gets shot… and we can't finish the game. The problem is, how do we decide who gets the $400? When the flipping stopped you are up two to one. How should the $400 be split? That's the problem of points.

In the seventeenth century, people would say that whoever is up 2–1 gets the $400, end of conversation. Because 375 years ago even the brightest minds thought, rather knew, that numbers cannot predict the future. Numbers were for counting. Or the seventeenth century minds would argue that the pot should be split 50–50 because nobody can determine God's will. Split the pot. But Pascal and Fermat started to think about what would happen, what could happen. This was a fundamental change in thinking; remember, God decided what happened. This gambling problem changed the world. And two of the greatest mathematicians of all time … got it wrong. They thought the "whoever is winning gets the $400" or "split the pot" approach was logical until they thought more deeply and discussed the problem – and created statistics.

Pascal and Fermat began to think about what could happen in the game, what was most likely to happen. This was blasphemous. Only God knew. But they realized four things could happen in the game: the coin could come up heads-heads (HH), heads-tails (HT), tails-heads (TH), and tails-tails (TT). Three of the four translate to you winning $400 and one of the four means

---

[4] Peter L. Bernstein. *Against the gods: The remarkable story of risk.* New York: Wiley, 1996.

that I win the $400. So, modern logic or expected value theory argues that the pot should be split $300 for you and $100 for me. The problem of points seems simple. But two legendary mathematicians struggled and debated and got it wrong over and over again in a famous series of letters. These letters became the foundation of modern-day statistics. Pascal and Fermat invented odds.

Odds are defined as occurrences versus non-occurrences. The odds of flipping heads on a coin flip is one to one or 1:1 or 50–50 or a 50% chance of landing on heads. The odds of picking a spade out of a deck of cards is 1:3 or one spade versus three hearts-clubs-diamonds or a 25% chance of picking a spade. There are 38 pockets where a ball can land on a roulette wheel: 18 red, 18 black, and 2 green. You can bet on the ball landing on the reds or blacks, but you cannot bet on green. Wesley Snipes quipped in *Passenger 57* that you should "always bet on black." Wesley was wrong.

Always betting on black leads to the casino taking your money. Black wins 18/38 or 47% of the time; red wins 18/38 or 47% of the time, and green wins 2/36 or the remaining, roughly, 6% of the time. The casino breaks even on the red and black bets while slowly picking up the green bets as you slowly lose. 47% means that the odds are against you Wesley, but gambling losses are tax deductible to the extent of any winning. One person did beat the roulette wheel: Edward O. Thorp.

Ed Thorp was a young MIT professor when he bought a roulette wheel which he used to study the ball's speed relative to its wheel position and where the ball landed. Then, he developed formulas that improved his chances of guessing where the ball would fall thus changing the odds of the game. Determined to beat the casino, he invented the world's first wearable computer in 1961 which he used inside casinos with some success.[5] So maybe you can beat the casinos, but only until they change the rules. You cannot bring a cell phone to the tables these days, thanks in some part to Ed Thorp's genius.

Independence and Coin Flips

On May 8th 2019, the Baltimore Orioles hosted the Boston Red Sox at Camden Yards. The Red Sox's Chris Sale threw nine straight strikes in the seventh inning and struck out three straight Orioles hitters—an immaculate inning. Twenty-eight days later on June 5, 2019, the Kansas City Royals hosted the Red Sox at Kauffman Stadium. Chris Sale threw nine straight

---

[5] Thorp, Edward O., *A Man for All Markets*. New York: Random House, 2017.

strikes in the eighth inning and struck out three straight Royals hitters—an immaculate inning. Sale threw two immaculate innings in 4 weeks.

Lefty Grove is the only other pitcher with two immaculate innings in the same season. Five other pitchers have thrown two immaculate innings: Sandy Koufax, Nolan Ryan, Randy Johnson, Max Scherzer, and Kevin Gausman. Sandy Koufax is the only player to pitch three immaculate innings. Since 1889, the immaculate inning happened 100 times in Major League Baseball. Like immaculate conceptions, immaculate innings are extremely rare. But this is changing, and improved umpiring is driving the change.

All humans, including umpires, suffer from the gambler's fallacy, meaning that we get fooled by random patterns.[6] For example: a gentleman flips a coin and gets two heads in a row, your brain may believe that the next flip will be tails, tails are "due." Or your brain may be convinced that the next flip will be heads again because heads are on a "hot streak." Either way, you're suffering from the gambler's fallacy. Regardless of the previous flips, the odds on the next flip are still 50–50. Coin flips are independent.

Suppose a gentleman flips a coin and it comes up heads ten times in a row. Then, he turns to you and says: "Let's bet $100 on the next flip, heads or tails?" Most people will reply "tails" because ten heads in a row demands that a tail must come. But coins have no memory. Regardless of the last five or ten or ten-million coin flips, the odds on the next coin flip are still 50–50. Humans, however, are hardwired to find patterns where none exist. So, ten heads in a row usually leads to a bet on tails; or ten reds on the roulette wheel leads to a bet on black; or a slot machine that hasn't "hit" in a while is "due," so ante up. But random processes like coin flips and roulette wheels are never "due." Coins do not have memories. Coin flips are independent. The gambler's fallacy means that we find bogus patterns in random events, underestimate the likelihood of streaks, and lose a lot of money at casinos.

If an umpire calls ten strikes in a row, the next pitch is likely to be called a ball regardless of the ball's location. Umpires develop an uneasiness with streaks of strikes, and 'expect' a ball. Bad calls ensue. The umps succumb to the gambler's fallacy. Wrapping one's mind around the idea of streaks of random events is difficult. Anxiety sets in. It's mentally tough to call ten strikes in a row even when the ball location demands those calls. But improved incentives are changing this dynamic.

---

[6] Chen, Daniel L., Tobias J. Moskowitz, and Kelly Shue. "Decision making under gambler's fallacy: Evidence from asylum judges, loan officers, and baseball umpires." *The Quarterly Journal of Economics* 131, no. 3 (2016): 1181–1242.

Umpires changed when "Zone Evaluation" (ZE) began in 2009. Three senior umpires got fired following the 2009 season because of poor ZE ratings. And those with positive ZE ratings received promotions and salaries that reached up to $400,000. Is job security and a $400,000 salary enough incentive to break the gambler's fallacy?

For 40 years from 1969 through 2008, the number of immaculate innings per year looks as follows[7]:

1969 through 2008–40 years		
Number of immaculate innings	Number of years	Percent of years (%)
0	17	43
1	13	33
2	3	8
3	5	13
4	1	3
5	1	3
6	0	0
7	0	0
8	0	0

Zero immaculate innings in 43% of the seasons and only one in another 33% of seasons. And here's how things change following the introduction of pitch tracking:

2009 through 2019–11 years		
Number of immaculate innings	Number of years	Percent of years (%)
0	0	0
1	2	18
2	5	45
3	0	0
4	1	9
5	0	0
6	0	0
7	1	9
8	2	18

No seasons without at least one immaculate inning, and the three highest immaculate innings totals by year in MLB history.

Pitch tracking causes an uptick in immaculate innings… probably. Eleven years is too short a window to draw strong conclusions. And other baseball norms have changed. The total number of strikeouts by year in the major leagues set a record in 2019 with 42,283 whiffs, passing the 2018 record of

---

[7] Baseball-Almanac.com

41,207 strikeouts. The total strikeout record has been broken successively for 14 years. Strikeouts seem to be ever increasing.

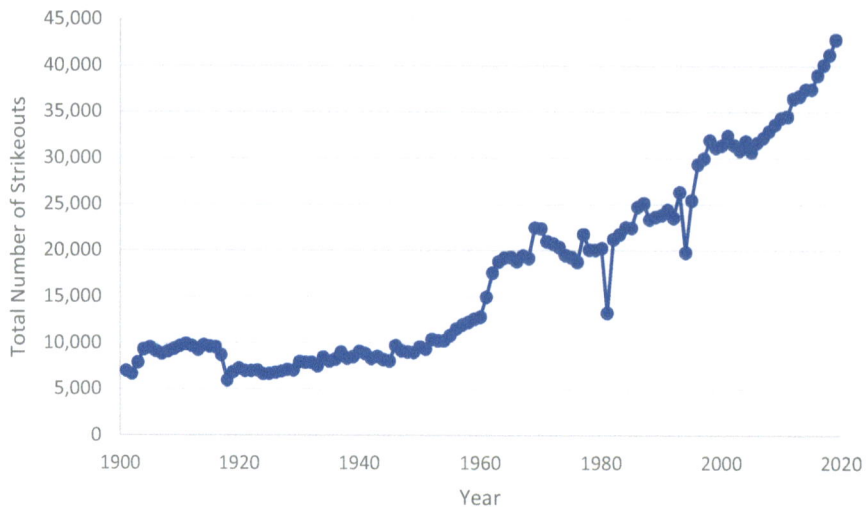

Strikeouts were not always sanguine. Joe DiMaggio and Mike Trout are the premier centerfielders of their time. Joe DiMaggio had 29 home runs and 39 strikeouts in his rookie year of 1936. Then, he vowed to Yankee fans that he would cut down on his strikeouts. Mike Trout won the rookie-of-the-year award in 2012 by swatting 30 home runs and striking out 139 times. He made no vows to anyone about cutting down on strikeouts. Joe DiMaggio earned his first MVP award in 1939 with 30 home runs and 20 strikeouts. Mike Trout earned his first MVP award in 2014 with 36 home runs and 184 strikeouts. Joe DiMaggio finished his Hall of Fame career after 13 seasons with 6821 at bats and 369 strikeouts. Mike Trout is 9 years into his Hall of Fame career with 4340 at bats and 1118 strikeouts. Times have changed.

So maybe pitch tracking has nothing to do with an uptick in immaculate innings. Maybe we should blame the increase on strikeouts and give the umpires a break. Or maybe both factors matter. Or maybe there is some other explanation out there. We can never know for sure, but we can make better bets.

Again, a gentleman flips a coin, gets ten straight heads, and asks for a bet of "heads or tails." What should you do? Those suffering from the gambler's fallacy will bet tails because a tail is "due," and coins have memories. Statisticians will explain that coin flips have no memory, so the odds are still even. But the best answer comes from Poker Hall of Famer Barry Greenstein. He writes in his book, *Ace on the River*, that a professional gambler would deduce, "There must be something wrong with the coin or the way it is being flipped. I wouldn't bet with the guy flipping it, but I'd bet someone else that heads will come up again."[8]

Tragedy and Coin Flips

In September 1996, Sally Clark was a 32-year-old new mother in the city of Chelmsford in Essex, England. Her infant boy, Christopher, died 11 weeks after his birth in December 1996. Two years later, Sally Clark bore a second child, Harry. Harry died in January 1998 8 weeks after his birth under similar circumstances. The police arrested Sally Clark in February 1998. They charged her with murdering her sons, Christopher and Harry.

During the trial, the prosecution argued that the probability of a child dying of sudden infant death syndrome or SIDS in an affluent family of non-smokers was 1 in 8573. Thus, the chance of two children both dying of SIDS would be less than 1 in 73 million $\left(\frac{1}{8,573} \times \frac{1}{8,573} = \frac{1}{73,496,329} \right)$. This mathematical logic—multiplying two probabilities together—is correct when the two events are 'independent,' like coin flips. The probability of flipping two heads in a row is 25% $\left(\frac{1}{2} \times \frac{1}{2} = \frac{1}{4} \right)$. The probability of birthing two baby boys in a row, also independent and also 25%. However, if the events are not independent, such as two brothers suffering from the same genetically transferred disease that causes SIDS, then the logic is horrifically wrong. That did not stop the prosecution from hammering these numbers down the jury's throat during the trial. "One sudden infant death is a tragedy, two is suspicious, and three is murder," testified Professor Roy Meadow.[9]

In November 1999, Sally Clark was convicted of murder. Sally Clark was sentenced to 2 life sentences. Sally Clark was sent to prison. Sally Clark was vilified in the press. Sally Clark was denigrated by her fellow inmates who could not comprehend a mother killing her own children. But her husband, Steven Clark, fought to have her conviction overturned and forced appeals where he received support from The Royal Statistical Society (RSS). The RSS

---

[8] Greenstein, Barry. *Ace on the River*. Last Knight, 2005.

[9] Meadow, Roy. 'Fatal Abuse and Smothering' in Meadow, R. (ed.), *ABC of Child Abuse*. BMJ Publishing, London, 1997.

is a British organization designed to promote the proper use of statistics.[10] Peter Green, the president of the RSS, released a public statement before the second appeal explaining the errors in the statistics used in Sally Clark's trial. He highlighted the erroneous independence assumption inherent in the 1 in 73 million number:

> *The calculation leading to 1 in 73 million is invalid. It would only be valid if SIDS cases arose independently within families, an assumption that would need to be justified empirically. Not only was no such empirical justification provided in the case, but there are very strong reasons for supposing that the assumption is false. There may well be unknown genetic or environmental factors that predispose families to SIDS, so that a second case with the family becomes much more likely than would be a case in another, apparently similar, family.*[11]

Professor Peter Green

Professor Green also detailed problems with the presentation. He argued that some jurors likely misconstrued the 1 in 73 million number as a probability of innocence. Lawyers routinely confuse juries with statistics. So much so that academics named these lies the "prosecutor's fallacy." In addition, Dr. Green questioned the accuracy of the 1 in 8573 number. The numbers in the trial were dubious and flimsy. The numbers were wrong.

Besides the debunking of the faulty statistics, medical evidence revealed that Harry had "colonization of staphylococcus aureus bacteria" which indicated that he died from natural causes. Sally Clark won her second appeal in January of 2003 and was released from prison after more than 3 years behind bars. But Sally never recovered. Who could? Sally died in 2007 at the tender age of 42 from acute alcohol intoxication. Christopher and Harry would have been 10 and 8 years old. Rest in peace Sally. Rest in peace Christopher. Rest in peace Harry.

Anxiety and Coin Flips

"Ref, you suck!" is the first episode of best-selling author Michael Lewis' podcast *Against the Rules*. Lewis visits the NBA's instant replay facilities in Secaucus, New Jersey and details the process the NBA uses to institute instant replay. Lewis points out two contradictory facts. First: referees are better than

---

[10] Royal Statistical Society Goals

[11] Schneps, Leila, and Coralie Colmez. *Math on trial: how numbers get used and abused in the courtroom.* Basic Books (AZ), 2013.

ever at making accurate calls. Second: fans, players, coaches, and journalists are more brutal than ever at berating referees.

Lewis argues that we're all upset about fairness. "Then one day you wake up in a world that seems not just unfair but actually sort of rigged. That is, it's incapable of becoming fair because the people who benefit from the unfairness have the power to preserve it."[12] He follows this with audio clips of President Trump and Senator Sanders. Both politicians rally their faithful with rhetoric about rigged systems and unfairness. But Lewis' logic is questionable. The world is unfair, yes. And every politician in the world for hundreds of years has used the "unfairness" cliché. The world has been unfair a long time. And basketball players have been complaining about referees for a long time. Unfairness disturbs people, sure, but it is uncertainty that raises blood pressures.

"We are prone to overestimate how much we understand about the world and underestimate the role of chance in events," wrote Nobel Laureate Daniel Kahneman in his best-selling book *Thinking, Fast and Slow*.[13] The world makes sense to us because our minds construct narratives that make sense to us. Not because we actually understand the world. Nobody understands basketball's charging versus blocking rule. Nobody. It's random. And everyone believes that their team is on the wrong side of that coin flip.

Foul calls in the NBA—and every sport—are subjective, random. Charging, blocking, travelling, goaltending, palming, and flagrants are all subjective. Instant replay cannot help. But can instant replay determine the more definitive calls: was his toe on the three-point line? Did the shot get off before the buzzer? Who touched the ball last? The answer is no. These calls are easier to determine with instant replay… most of the time. But not when they matter. The sad situation is that instant replay cannot get them "correct" either. The close calls come down to judgment no matter what. The world is random. The world is uncertain.

Every modern-day sports fan is flummoxed by countless angles and slower than slow-motion, two-dimensional images of a foot and a sideline shown over and over until said fan screams for the good old days when the ref made the call and the game moved on. Instant replay does not make a subjective decision into a definitive decision any more than lipstick changes the molecular makeup of a pig. The number of "angles," the slowness of the images, the high definition, and the ex-referee explaining the nonsense cannot make a subjective decision definitive. And sports fans do not want to watch a ball

---

[12] "Ref, you suck," *Against the Rules* podcast, Episode 1, Michael Lewis.
[13] Kahneman, Daniel. *Thinking, fast and slow*. Macmillan, 2011.

leaving a player's fingers 50 times to guess if the shot clock expired. Play the game please. So, why do we have instant replay? Why has every major sport across the globe adopted instant replay in some way, shape, or form? And why do they call it instant when it takes forever?

Kahneman's lesson: we are prone to underestimate the role of chance in events. We do not like chance. We believe that we can control things that we cannot control. This is the human condition. So, when the referee calls a charge versus a blocking foul, the idea is not that the call is subjective, and that chance plays a role. The reaction is that the referee blew the call. And that's unfair for my team. We fail to acknowledge uncertainty. And failing to acknowledge uncertainty goes far beyond sports.

On October 31st 2019, the Bureau of Labor Statistics (BLS) sent a news release that stated that the "total nonfarm payroll employment rose by 128,000 in October. All the major news agencies reported this "fact." Some reporters even drew dubious conclusions that the stock market rallied because of the report. Not to be outdone, President Trump jumped on the dubious conclusions bandwagon with the following tweet:

> Wow, a blowout JOBS number just out, adjusted for revisions and the General Motors strike, 303,000. This is far greater than expectations. USA ROCKS! – Donald J. Trump (@realDonaldTrum) November 1, 2019

Unfortunately for the dubious concluders, the BLS news release also said:

> This news release presents statistics from two monthly surveys. The household survey measures labor force status, including unemployment, by demographic characteristics. The establishment survey measures nonfarm employment, hours, and earnings by industry. For more information about the concepts and statistical methodology used in these two surveys, see the Technical Note.

"See the technical note." The technical note explains the statistical realities about the uncertainty of estimates in a sample. Think of this as if your spouse asks when you will be home and you reply with "between thirty minutes and an hour." You both know that there is no certainty with rush hour traffic. The BLS answers their spouse with:

> For example, the confidence interval for the monthly change in total nonfarm employment from the establishment survey is on the order of plus or minus 110,000. Suppose the estimate of nonfarm employment increases by 50,000 from one month to the next. The 90-percent confidence interval on the monthly change would range from -60,000 to +160,000 (50,000 ± 110,000).

Trump's 128,000 job increase falls between 14,000 jobs and 238,000 jobs but we are not sure where. And 10% of the time even that range is too small. Maybe the jobs numbers are even negative. And maybe it will take an hour and a half to get home on an unlucky night. The uncertainty is due to sampling variability.

So, what can you conclude from the 128,000 jobs announcement? Absolutely nothing. Nothing. Nothing about the economy. Nothing about the job the president is doing. Nothing about the stock market. Nothing. Make early 1970s Edwin Star and late 1980s Bruce Springsteen and middle 1990s Elaine Benes proud and say it again: absolutely nothing. Certainty is an illusion. The same chef, using the same recipe, cannot bake the same cake twice. You will get randomness. How the ingredients are mixed will be slightly different. The temperature and time the cake is cooked will change. Innumerable things will be altered. You cannot bake the same cake twice. Uncertainty is the rule.

Acknowledging uncertainty is difficult. "Cartesian Anxiety" is a term connected to the seventeenth century philosopher René Descartes that refers to a longing for absolute certainty. The belief that science should lead us to a firm and unchanging knowledge of ourselves and the world around us. But science cannot provide that fantasy. The world is too complicated. Uncertainty rules.

Life would be easier if the all-knowing referee made the "correct" call every time. Everything would be "fair." Michael Lewis would sleep better. Reporters and the president would be happier if we could count the quarterly job changes. But we cannot. We also cannot get charging calls "correct" in basketball games or predict the traffic on our way home from work or bake the same cake twice or predict any of the important things in life—success, health, friendship, love. The eminent mathematician John von Neumann remarked: "Truth … is much too complicated to allow for anything but approximation."

Uncertainty is reality. But that may not be bad. Uncertainty is the straw that stirs the drink, the tequila in your margarita. Take away uncertainty and we're left with a robotic existence, no free will, and no reason to exist. Also, no last second buzzer beaters where the underdog knocks out the favorite. Uncertainty is life. So, let's not just acknowledge uncertainty, let's embrace uncertainty. After all, what's fairer than a coin flip?

3

First Principles

Detective Work

In the 1991 best picture *The Silence of the Lambs*, Hannibal "the Cannibal" Lecter chides aspiring FBI agent Clarice Starling: "First principles, Clarice. Simplicity. Read Marcus Aurelius. Of each particular thing ask: what is it in itself? What is its nature? What does he do, this man you seek?" Lecter scolds Clarice for not understanding the teachings of stoic philosophy and the work of Marcus Aurelius—a Roman emperor from 161 AD to 180 AD. First principles thinking teaches breaking down problems into the most basic elements. Looking at problems from different angles. Avoiding bias. Asking what makes up the problem, what patterns exist.[1] Good detectives solve cases this way. And good statisticians practice first principles too.

Statistics is detective work. Making sense of numbers is not mathematics. Mathematics involves following established rules in fixed settings such that two plus two always equals four. Mathematics is logic. Statistics is judgment. Two plus two kind of equals four… sometimes. Maybe. Statistics simplify. They compress reality. They omit information. And this is both a feature and a bug. Statistics summarize numbers making them capable of being understood, digested. But summarizing means simplifying, which can lead to misinformation and mistakes. When done well, statistics illuminate the hidden and the unseen. When done poorly, statistics mislead and confuse. Statistics is art. Statistics is reasoning. Statistics is detective work.

---

[1] Aurelius, Marcus. *Meditations*. Penguin, 2015.

R. T. Stewart, *Adventures in Statistics*, Copernicus Books ,
https://doi.org/10.1007/978-3-031-61284-8_3

If you sum the heights of everyone in the movie theatre watching *The Silence of the Lambs* and divide by the number of people, you will compute the average or the mean height.[2] Averages contain two pieces of information: the total and the number contributing to the total. But if basketball hall of famer Shaquille O'Neil and some of his former Los Angeles Lakers cohorts drop by the theater that day, your average calculation will mislead. Shaquille's 7 feet, 1 inch height along with his fellow basketball friends' heights will pull the average up. Your average is biased.

Never trust an individual statistic. If you're given an average, ask for a median. If you line up everyone in the movie theatre by height and take the middle person, that person's height is the median. One observation, uninfluenced by the Laker outliers in the lineup. First principles include looking at problems from different angles, avoiding bias. Averages and medians go together like Tango dancers. Two pieces of evidence are more valuable as a pair. Averages give you total values; medians give you typical values. Averages are sensitive to outliers; medians are indifferent to outliers. You must use all the evidence to make sound decisions, Clarice.

In the penultimate scene of *The Silence of the Lambs*, Clarice Starling calls Jack Crawford, her boss, to report her conversation with Lecter. Jack cuts Clarice off mid-sentence to exclaim that they know the killer's identity. His name is Jamie Gumb. He was rejected by Johns Hopkins for transsexual surgery, has a criminal history, and a customs bill of lading for moth pupae is addressed to him. Jack Crawford is on a plane to Chicago to save the girl and arrest the murderer. But Jack gets it wrong. And Jack puts Clarice in danger. Always question and be cautious when making sense of averages or any pattern in data or statistics. Lest you, like Jack Crawford, end up chasing ghosts.

The Normal Distribution

Madison Square Garden in New York City seats 20,789 people at capacity. Dubbed "The World's Most Famous Arena," Madison Square Garden has provided the backdrop for countless earth-shattering performances including those of Elvis Presley, Muhammad Ali, Marilyn Monroe, Michael Jordan, Bruce Springsteen, and Pope Francis. After Elvis shook his hips in Madison Square Garden, music changed. Rock and roll exploded across the world. The

---

[2] Averages and means are identical. The same formula is used to compute averages and means. The words are used interchangeably by statisticians.

normal distribution or the bell curve is the Elvis of statistics; the normal distribution revolutionized statistics.

The normal distribution began blandly as a mathematical treatise about binomial probabilities in the eighteenth century. But the pattern described by the normal distribution, the bell curve, keeps on showing up. It cannot be stopped. The best visual for the normal distribution is height, which most statisticians feel comfortable describing as a normal distribution. Two numbers or parameters describe the normal distribution: the average and the standard deviation. The average is the sum of a group of people's heights divided by the number of people measured. The standard deviation is more interesting. This measures the spread or range of the heights.

Suppose we go to Madison Square Garden to enjoy a Bruce Springsteen show, and measure the heights of all the males who enter the building. Given that Madison Square Garden holds roughly 20,000 people, we'll measure about 10,000 men. The sample will look something like this:

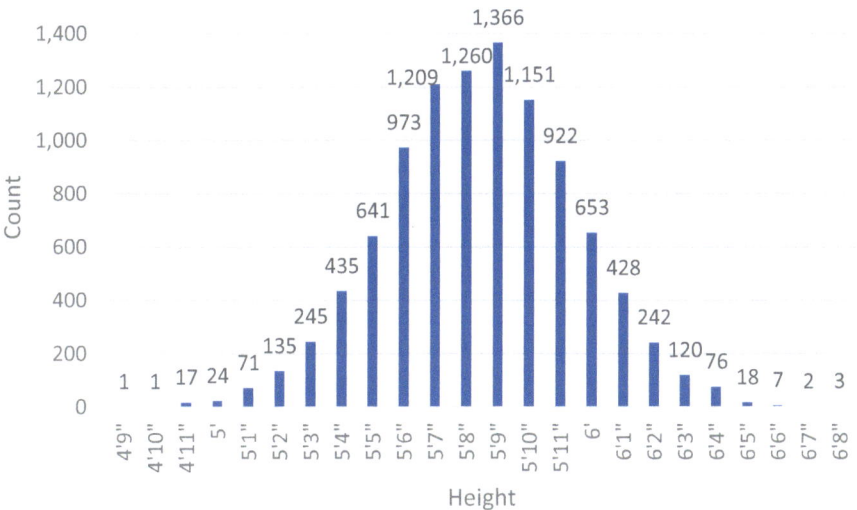

The above chart is a histogram which separates each person's height into a bucket. So, the first bucket on the far left is everyone who is 4 feet, 9 inches tall or one potential jockey in the sample. The next bucket is everyone who is 4 feet, 10 inches and so forth up to 6 feet, 8 inches—the tallest, potential NBA stars in the sample. The bucket with the largest count is the average of 5 feet, 9 inches where 1366 of the 10,000 in our sample fall.

Normal distributions look like bells, hence the name. And they are symmetrical, like bells, meaning that the same number of people are shorter than the average of 5 feet, 9 inches as are taller. But the interesting thing about the normal distribution is the standard deviation. The standard deviation describes the distance from the average across data points. So, if the average height is 5 feet, 9 inches and you're 5 feet, 5 inches, your distance from the average is 4 inches. The average distance from 5 feet, 9 inches tall for all the data points is the standard deviation. The standard deviation for male height is about 3 inches. Because height is normally distributed, 66% of the guys measured fall within one standard deviation of the average. Thus, roughly 6666 of the 10,000 measured men in our sample will be between 5 feet, 6 inches tall and 6 feet tall. Similarly, 95% of the heights will fall between two standard deviations or between 5 feet, 3 inches and 6 feet, 3 inches tall. That's our model of height… does it seem reasonable? Be cautious.

Average body temperature is 98.6 degrees Fahrenheit or 37 degrees Celsius. This temperature is healthy and "normal." Everyone knows this. The song "98.6" reached number 7 on the Billboard chart and sold over a million copies worldwide in 1967. Everyone knows that 98.6 is normal, it's in a hit song. But the 98.6 degrees Fahrenheit average body temperature that we "know" as healthy and "normal"… is wrong.

Three scientists wrote a paper in 1992 that concluded, "We believe that 37 degrees Celsius or 98.6 degrees Fahrenheit should be abandoned as a concept having any particular significance for the normal body temperature."[3] Oops. The authors found that 98.2 degrees Fahrenheit was a more accurate estimate. And that temperature should and would fluctuate for countless reasons including sleep patterns, sex, menstrual cycles, age, and race. The 98.6 number came from a German physician named Carl R.A. Wunderlich. He took the temperatures of more than 25,000 people over 16 years before publishing a paper in 1868. The thermometers in 1868 were inaccurate and the place where he took the temperature—armpit, mouth, and rectum—were not consistent. The 98.6 degree finding is wrong. Still, the 98.6 number stuck around for 150 plus years and counting.

The Washington Post quoted one of the scientists who wrote the paper: "The whole idea should be abandoned that a single temperature has any significance as the "normal" temperature. One should think of the 'normal' temperature as a range, rather than any particular number."[4] Averages don't tell us much. Having a temperature a little lower or a little higher than the average is

[3] Mackowiak PA, Wasserman SS, Levine MM. A Critical Appraisal of 98.6 degrees Fahrenheit, the Upper Limit of the Normal Body Temperature, and Other Legacies of Carl Reinhold August Wunderlich. JAMA. 1992; 268 (12): 1578–1580.

[4] Brown, David. "Normal body temperature found to be 98.2 degrees." *Washington Post*, 23 September 1992.

not unhealthy. We should think of this as equal to being a little shorter or a little taller than the average height. Smart detectives explore the range around the averages, the standard deviations. First principles, Clarice.

Suppose Hannibal tells Clarice that he will pay her $1000 for each head she gets in ten coin flips. How much should Clarice pay to take that bet? $100? $1000? $5000? $10,000? Clarice can experiment to find out. She can flip a coin ten times, add up the heads; then, flip a coin ten more times and add up the heads… and do this 100,000 times or so. Luckily, Clarice's computer makes those calculations for her:

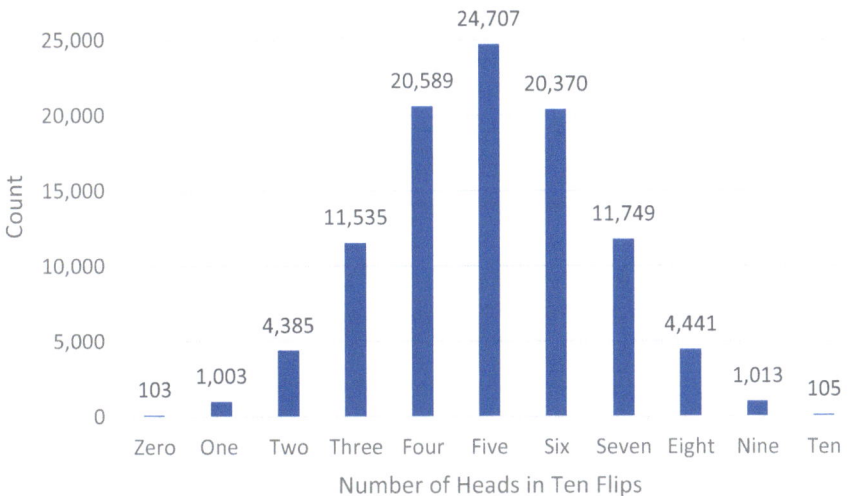

Now, how much should Clarice pay for $1000 per head on ten flips? There is a chance that she gets nothing, but there is also a chance that she makes $10,000. Roughly two-thirds of the realizations land at $4000; $5000; and $6000. The results look like a bell curve or a normal distribution. But a coin flip is random? Can there be a pattern?

Even with random processes such as coin flips, sums of random samples—such as sums of heads in ten coin flips—will follow a normal distribution. This phenomenon is known as the central limit theorem. Like the backbeat Elvis used to change music, the central limit theorem changed statistics. A pattern derived from a random process. Random processes can be tamed, read Marcus Aurelius.

The central limit theorem explains part of the reason that the normal distribution is ubiquitous in so much of statistics, but nature also plays a role. What would happen in a world where height was not distributed normally? Shaquille O'Neill stands 7 feet, 1 inch tall. According to our model of height as a normally distributed variable with an average height of 5 feet, 9 inches tall and a standard deviation of 3 inches, the Big Shamrock is more than 5 standard deviations or 15 inches taller than the average height. Shaq is taller than 99.99% of the population.

Shaq Daddy is a father. His oldest son, Shareef, plays college basketball and stands 6 feet, 9 inches tall. Like his dad, Shareef is taller than 99.99% of the population. But Shareef is shorter than Shaq. Shareef "regresses to the mean" or average. Tall people tend to have children shorter than they are; and short people tend to have children taller than them. Children trend toward the average. God set the world up this way because the alternative would not work. If Shaq had kids taller than him, and those kids had taller kids, and those kids had taller kids... before long, Shaq's ancestors would be 20 feet tall. The same idea with trees. Tall trees have saplings that lead to somewhat shorter trees, so nature "regresses to the mean" or average. God has a plan.

But Hollywood rules are different. The 1958 classic *Attack of the 50 Foot Woman* and the 1993 remake with the same name starring Daryl Hannah received no love from the Academy of Motion Pictures Arts and Sciences. Perhaps because a 50 foot woman would have some serious health issues due to gravity and physics. A woman whose height is increased ten times, must also have her width, and thickness increased by ten times to maintain proportionality. Increasing the length, width, and thickness each by ten times means that volume increases by ten times ten times ten—length times width times height. Mathematicians call this an exponential increase. So, the 5-foot woman who grows into a 50-foot woman is going to gain a lot of weight, much more than ten times the weight. More weight than human bones can stand. Human bones would collapse under the pressure of the exponential increase in weight. The 50-foot woman is impossible.

Large animals like rhinos, elephants, and hippopotami all have thick legs to support their weight. Daryl Hannah would need some serious thighs and she could not look like she does in the movie. But if Daryl Hannah could grow some gills like her character in *Splash* and live in the ocean as the 50-foot woman with gills, she might have a fighting chance. The buoyancy of the water can support the extra weight without crushing her bones. The blue whale, the largest animal on earth, is as long as a basketball court and weighs in at around 150 tons or about 20 African elephants, the largest animal on land. Because the blue whale lives in water, the 140 or so tons of weight it has over the elephant are more manageable. So, nature explains why Shaq's kids will "regress to the mean"… or grow gills.

"Regression to the mean" has further implications. Daniel Kahneman won the 2002 Nobel Prize in economics for applying psychological insights into economic reasoning. Kahneman describes 'regression to the mean' as follows:

> I had the most satisfying Eureka experience of my career while attempting to teach flight instructors that praise is more effective than punishment for promoting skill-learning. When I had finished my enthusiastic speech, one of the most seasoned instructors in the audience raised his hand and made his own short speech, which began by conceding that positive reinforcement might be good for the birds, but went on to deny that it was optimal for flight cadets. He said, "On many occasions I have praised flight cadets for clean execution of some aerobatic maneuver, and in general when they try it again, they do worse. On the other hand, I have often screamed at cadets for bad execution, and in general they do better the next time. So please don't tell us that reinforcement works and punishment does not, because the opposite is the case." This was a joyous moment, in which I understood an important truth about the world: because we tend to reward others when they do well and punish them when they do badly, and because there is regression to the mean, it is part of the human condition that we are statistically punished for rewarding others and rewarded for punishing them.[5]

Suppose you had Shaq and Shareef flip a coin ten times. Shaq gets 8 heads, so the coach praises his wonderful job while Shareef gets only 2 heads, so the coach criticizes his poor job. Then, they flip the coin another ten times. Shaq will likely get less than eight heads and Shareef will likely get more than two heads. So, an uninformed coach can believe that praise leads to bad outcomes while criticism leads to good outcomes.

Most tasks have an element of luck or randomness involved. You can't bake the same cake twice. Luck plays a role in flying jets, shooting free throws, and baking cakes. Luck plays a role in almost everything, so "regression to the mean" is pervasive. We underestimate the role of chance in our lives. Coaches of any kind should keep this in mind when delivering both criticism and praise. Shaq made free throws at a career 52.7% rate, like coin flips. But one must remain doubtful that the Big Diesel was much influenced by praise or criticism from his coaches. He attributes his free throw adventures to the Big Deity, saying: "So me shooting 40% at the foul line is just God's way of saying that nobody's perfect. If I shot 90% from the line, it wouldn't be right."[6] Good point.

[5] Kahneman, Daniel. *Thinking, fast and slow*. Macmillan, 2011.
[6] Gonzalez, AJ. (September 15, 2021) Shaquille O'Neal once made a hilarious comment about his free-throw shooting, *Sports Illustrated*.

Power Laws

While Shaq's confidence is admirable, hubris does not bode well for Shaq's budding career in criminal science. Jack Crawford would teach Shaq that good detectives must be open to different viewpoints and contrasting theories. Productive statisticians are like good detectives in that they should explore multiple viewpoints when reading the clues in the numbers, while understanding that red herrings abound. When we examine numbers from the view of a normal distribution, the average and standard deviation play central roles. But averages and standard deviations are not always meaningful.

In 2015, 158 movies were released in 600 theatres in the United States.[7] Suppose we're interested in domestic sales for those 158 movies. Let's start with looking at the top ten highest grossing movies in 2015:

Rank	Title	Total gross
1	Star Wars: The Force Awakens	$936,662,225
2	Jurassic World	$652,270,625
3	Avengers: Age of Ultron	$459,005,868
4	Inside Out	$356,461,711
5	Furious 7	$353,007,020
6	American Sniper	$350,126,372
7	Minions	$336,045,770
8	The Hunger Games: Mockingjay Part 2	$281,723,902
9	The Martian	$228,433,663
10	Cinderella (2015)	$201,151,353

And the bottom ten lowest grossing movies in 2015:

Rank	Title	Total gross
149	Truth	$2,541,854
150	Pawn Sacrifice	$2,246,000
151	My All American	$2,184,640
152	Jem and the Holograms	$1,650,000
153	Back to the Future Day	$1,650,000
154	The Letters	$1,647,416
155	The Diary of a Teenage Girl	$1,477,002
156	99 Homes	$1,411,927
157	Patterns of Evidence: The Exodus	$925,576
158	The D Train	$673,151

The difference between the top movie sales and the bottom movie sales are enormous. *Star Wars: The Force Awakens* is just under a billion dollars or 8% of the total industry sales for the year. And Black Jack's *The D Train* barely

---

[7] Numbers from boxofficemojo.com

broke six hundred thousand dollars. If this was height, *Star Wars* would be the 50-foot woman and Black Jack would be a jockey riding in the Preakness Stakes. A histogram of movie sales for the top 158 movies:

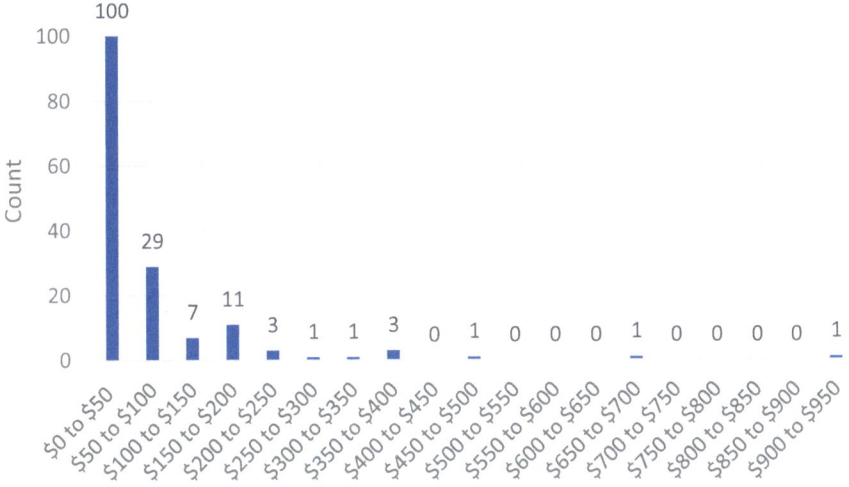

Total Domesitc Gross 2015 in Millions

The picture does not look like a bell curve; it looks like a mistake. There are 100 movies in the first bin on the left that made less than $50 million. Then, the second bin representing movies with sales between $50 and $100 million has 29 movies in it. So, more than 80% of all the movies in the sample fall inside the first two bins. They don't make any money.

But a handful of movies make enormous sales. The 11 movies that grossed more than $200 million make up 39% of total movie sales, and the 30 movies that grossed more than $100 million make up 64% of the total gross. Sales are top heavy. The average sales for the 158 movies is $70 million with a standard deviation of $115 million. But the average and the standard deviation do not mean much when the histogram is not symmetrical, not a normal distribution.

When the histogram looks ugly and nothing like a bell curve; and the standard deviation is well above the average; and your largest number is dramatically bigger than your smallest number, then you are not looking at a normal distribution. You're dealing with a fat-tailed or long-tailed distribution and possibly what's referred to as a power law. There's an aphorism that states: "If all you have is a hammer, everything looks like a nail." The detective who only searches for the usual suspects will fail. The statistician who always relies on the normal distribution will also fail.

Testing for power laws, and thus shifting attention from the average and standard deviation paradigm, allows shrewd detectives to find meaningful patterns in data. One tool is graphing the ranks of the data. For movies sales, the largest sales are dramatically larger than the smallest sales. Let's search for a pattern. First, we rank the 29 movies that had sales greater than $100 million.

Rank	Title	Total gross
1	Star Wars: The Force Awakens	$936,662,225
2	Jurassic World	$652,270,625
3	Avengers: Age of Ultron	$459,005,868
4	Inside Out	$356,461,711
5	Furious 7	$353,007,020
6	American Sniper	$350,126,372
7	Minions	$336,045,770
8	The Hunger Games: Mockingjay Part 2	$281,723,902
9	The Martian	$228,433,663
10	Cinderella (2015)	$201,151,353
11	Spectre	$200,074,609
12	Mission: Impossible–Rogue Nation	$195,042,377
13	Pitch Perfect 2	$184,296,230
14	Ant-Man	$180,202,163
15	Home (2015)	$177,397,510
16	Hotel Transylvania 2	$169,700,110
17	Fifty Shades of Grey	$166,167,230
18	The SpongeBob Movie: Sponge Out	$162,994,032
19	Straight Outta Compton	$161,197,785
20	San Andreas	$155,190,832
21	Mad Max: Fury Road	$153,636,354
22	Daddy's Home	$150,357,137
23	The Divergent Series: Insurgent	$130,179,072
24	The Peanuts Movie	$130,178,411
25	Kingsman: The Secret Service	$128,261,724
26	The Good Dinosaur	$123,087,120
27	Spy	$110,825,712
28	Trainwreck	$110,212,700
29	Creed	$109,767,581

Then, we use a little mathematical trick to find a pattern. Taking the logarithm of the rank and the logarithm of the sales diminishes the dramatic differences between the sales figures allowing for clearer comparisons. The graph of the logarithm of the rank and the logarithm of the sales:

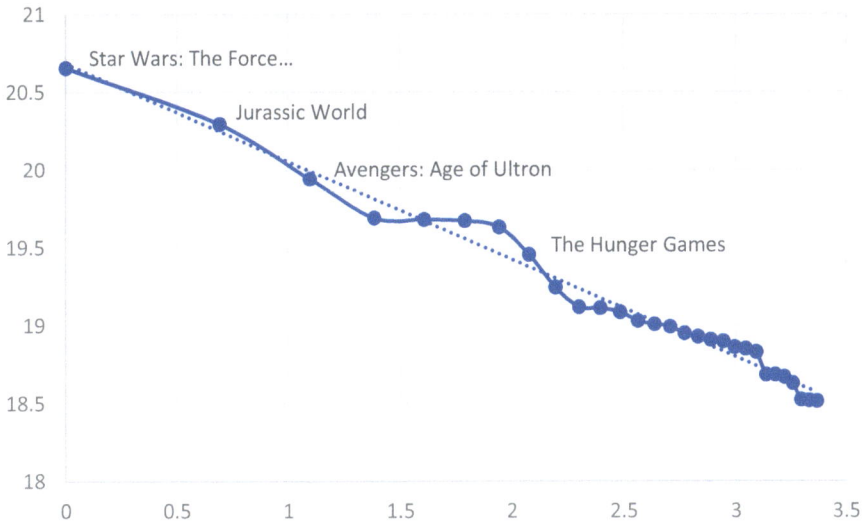

This statistical trick creates a straight line, a recognizable pattern. This line allows us to make predictions and improve decision making. If we produce five superhero movies and four flop, but one becomes *The Avengers* with almost half a billion in sales, we've got a profitable business. The 80% flop rate is fine even with the large costs involved in making a superhero movie. Most movies flop. But some make a fortune. The result is that Hollywood produces a lot of superhero movies.

The straight line, linear relationship between the logarithm of the rank and the associated logarithm of the sales is an example of the power law. This pattern shows up in a lot of places. Sales of most books, songs, Broadway shows, and shoe styles are small. A few huge books like Harry Potter, and a few huge singers like Taylor Swift, and a few huge Broadway shows like Hamilton, explain most of the total sales in each genre. Meghan Markle sports some Aquazzura Casablanca multi-strap suede pumps … and sales go crazy, but most shoes sales are mundane. Power law relationships arise when whatever is being measured—shoe sales—are not independent. When Meghan Markle wears the shoes in public a feedback loop is created as Meghan's fans buy the shoes. When an 8-year-old tells classmates that *Harry Potter and the Sorcerer's Stone* is an awesome book, then a feedback loop is created when the kids tell more and more kids. In these cases, an action—such as Meghan wearing a certain product or a child buying a book—increases the likelihood that more people will do the same. These feedbacks and interdependencies cause the power law. Contrast that with something like height which is independent. You cannot change your height based on Meghan Markle's whims.

Researchers have found the power law relationship all over. "Word frequency" follows a power law with "the' and "a" acting as the *Star Wars* of the written world. Web "hits" defined as when a page is downloaded follow a power law with YouTube, Google, Facebook, and Amazon dominating. Net worth follows a power law with the Bill Gates and Warren Buffets and Jeff Bezos owning billions while everyone else ends up in that left bucket. Researchers have found power law in citations of scientific papers, magnitudes of earthquakes, intensities of wars, city sizes, firm sizes, and stock market movements… to name a few.[8] The more you think about the power law, the more you recognize it all around you. Look for it, look for interdependencies and feedback loops, do not ignore it.

First Principles

Robert Pershing Wadlow is the tallest person in recorded history. Known as 'The Gentle Giant,' Wadlow suffered from a pituitary gland problem which resulted in abnormal growth. Born in Alton, Illinois on February 2nd 1918, Wadlow grew taller than his father at 8 years old when he hit about 6 feet. At 12, he was 7 feet tall and towered over his classmates. He died at the age of 22 after a faulty leg brace led to an infection. The infection combined with an autoimmune disorder, and killed him in his sleep. At the time of his death, he stood 8 feet, 11 inches tall. If we assume that height is normally distributed with an average of 5 feet, 9 inches tall and a standard deviation of 3 inches, then Robert Wadlow was 12 ½ standard deviations or 3 feet, 2 inches above the average. The tallest person in history.

David Alan Viniar was the Chief Financial Officer (CFO) at Goldman Sachs from 1999 through 2013. On August 13th, 2007, The Financial Times quoted Viniar, "We were seeing things that were 25 standard deviation moves, several days in a row."[9] He was explaining why two hedge funds managed by Goldman Sachs lost billions of dollars after dramatic changes in asset prices. Statisticians and the press mocked Viniar's statement.[10] According to our height model, 25 standard deviations would make you 12 feet tall. That's 4 feet taller than Robert Pershing Wadlow. But standard deviations are meaningless when what you're analyzing resemble a power law rather than a normal

---

[8] Gabaix, Xavier. "Power laws in economics: An introduction." *Journal of Economic Perspectives* 30, no. 1 (2016): 185–206.

[9] Reported in the Financial Times, August 13, 2007.

[10] Dowd, Kevin, John Cotter, Chris Humphrey, and Margaret Woods. "How unlucky is 25-sigma?." *The Journal of Portfolio Management* 34, no. 4 (2008): 76–80.

distribution. Viniar made the mistake of modeling asset prices as if they were height as opposed to movie sales. There will never be a 50-foot woman, but a movie that grosses $10 billion is feasible. Asset prices resemble power laws more than normal distributions. Asset prices are interdependent and subject to feedback loops. First principles, Clarice. What is its nature?

When investigating data, statisticians must look at averages and medians and standard deviations and histograms. They must think about the nature of the data which means asking questions without specific answers: *Does my data look like a normal distribution? Is it like height? Does my data look like a power law? Is it like movie sales? Is the data independent or interdependent? Are there feedback loops? What makes my data look the way it does? What are the underlying processes that are driving the numbers?* This is where your detective work will begin, but not where it ends. You must go further if you want the lambs to stop screaming.

4

Black Swans

Black Swans

Titanic won the Academy Award for Best Picture in 1997. The movie portrays the sinking of the maiden voyage of the RMS (Royal Mail Service) Titanic in the North Atlantic in 1912. The RMS Titanic was the biggest ship afloat at the time and thought to be "unsinkable." Then, more than 1500 passengers and crewmembers died when the unsinkable sunk. The movie received incredible hype before the opening and ran away with 11 academy award nominations due mainly to the groundbreaking special effects. But as the movie has aged and the cutting-edge special effects look more mundane each year, folks are less enamored with the stale, one-dimensional characters. Caledon "Cal" Nathan Hockley plays the snobby, rude, spoiled, selfish cliché of the rich kid. Cal declares before boarding the ship, "It is unsinkable. God himself could not sink this ship!" Many statisticians consider the sinking of the Titanic to be a "black swan." The term black swan means a surprise event that has a major impact and is often rationalized after the fact. New York University professor Nassim Taleb popularized black swan theory through a series of papers and books. He argues that black swans have three attributes: unpredictability, consequences, and retrospective explainability, like the Titanic.[1]

Legend says that the expression "black swan" was first used by Aristotle around 300 BC to explain the improbable, showed up in the poems of Juvenal

---

[1] Taleb, Nassim. *Fooled by randomness: The hidden role of chance in life and in the markets.* Vol. 1. Random House Incorporated, 2005.

Taleb, Nassim Nicholas. *The black swan: The impact of the highly improbable.* Vol. 2. Random House, 2007.

© The Author(s), under exclusive license to Springer Nature Switzerland AG 2024

R. T. Stewart, *Adventures in Statistics*, Copernicus Books ,

https://doi.org/10.1007/978-3-031-61284-8_4

in 82 AD to describe a creature that did not exist, and was commonly used in London in the sixteenth and seventeenth centuries to describe anything that was impossible or non-existent. All the swans in Europe were white, so the term made sense in the way that we may say "a snowball's chance in hell" today. Then, everything changed. A Dutch explorer named Vlamingh ventured to Western Australia in 1867 near modern day Perth and saw... a black swan. The news was shocking, this would be like someone in today's world traveling to hell and having a snowball fight. Then, the idiom "black swan" meaning impossible, changed to the idiom "black swan" meaning something unpredictable and surprising.

Black swans provide an invaluable lesson: statistics and history provide certainty of nothing. Europeans saw thousands of white swans over thousands of years. But no matter how many swans or how many years, empirical data cannot prove the absence of black swans. No matter how many studies or how many samples, statistics cannot prove that black swans exist or that black swans do not exist. Nassim likes to use the turkey example to make this point: a turkey using experience and data may conclude that the farmer is wonderful. The turkey may use fancy regression models and machine learning techniques that show how wonderful the farmer is: the farmer makes sure that the turkey eats and gets fat; the farmer protects the turkey from predators and provides a nice home. Then, Thanksgiving rolls around.

Statistics have limitations. Statisticians, like detectives, must always be wary of unknown unknowns. This unromantic notion bothers our human need for patterns, explanations, understanding, and answers. Humans love recognizing patterns because they make us feel comfortable and in control. Admitting that we don't know is like nails on the chalkboard. We'd rather make things up. We'd rather find patterns where they don't exist than admit that we don't understand the pattern or—horror—that the pattern is random. Acknowledging uncertainty is difficult.[2] "Cartesian anxiety" is a term connected to the seventeenth century philosopher René Descartes that refers to a longing for absolute certainty, and the belief that science should be able to lead us to a firm and unchanging knowledge of ourselves and the world around us. But uncertainty is the rule in a complicated world regardless of our desire to connect the dots.

My team will win as long as I wear my lucky hat. Don't walk underneath that ladder. Knock on wood. Some researchers postulate that pattern recognition is evolutionary, stemming from our need to react to dangers as hunter gatherers. Therefore, humans are excellent at finding patterns, but not so great

---

[2] Damasio, Antonio R. *Self comes to mind: Constructing the conscious brain*. Vintage, 2012.

at determining which patterns are meaningful.[3] We evolved such that humans who heard footsteps, and recognized that a lion was stalking, tended to live and procreate. While those who failed to recognize the pattern didn't get their genes passed down. So, humans are good at recognizing patterns. But humans are not good at understanding the credibility of their pattern recognition skills such as realizing that the footsteps they heard came from a chipmunk, not a lion. Hence, we are confident in our abilities to find patterns, too confident. The game is starting, where is my lucky hat? Hubris must be avoided in statistical analysis and black swans must always be respected lest you end up like young Cal Hockley watching the unsinkable sink.

Sand Piles

Some systems are prone to black swans. A Danish physicist named Per Bak imagined a game where you drop one grain of sand every turn onto a flat surface at the same spot.[4] The majority of time, the sand pile game is boring. One grain of sand is dropped on the pile and nothing happens. But as the pile grows, the likelihood of an avalanche where mounds of sand move around haphazardly increases, a usually insignificant event—dropping a grain of sand—causes remarkable change and a game that was seemingly stable and boring suddenly and unpredictably turns to bedlam. The size and timing of these avalanches are impossible to predict. We cannot know what grain of sand will cause the chaos.

What things in real life act like Per Bak's metaphor, the sand pile game? Can small, seemingly insignificant changes lead to large, crazy changes? Are there avalanches out there waiting? The power grid works great, and everyone takes it for granted, but on August 14th 2003, power was knocked out across the eastern United States and parts of Canada with 50 million people affected. The outage shut down trains, elevators, and subways; hospital services were interrupted, refrigerated foods were spoiled, and cellular phone services stopped. Power was restored in a day, but investigators were confused as to what caused the problem. What grain of sand caused the immensely complicated grid to break down? The investigation found that a small Ohio company shut down unexpectedly because overgrown trees touched the power lines.[5]

---

[3] Foster, Keving R., and Hanna Kokko. "The evolution of superstitious and superstition-like behaviour." *Proceeding of the Royal Society B: Biological Sciences* 276, no. 1654 (2008): 31–37.

[4] Bak, Per. *How nature works: the science of self-organized criticality.* Springer Science & Business Media, 2013.

[5] Taylor, Alan. "Photos: 15 Years since the 2003 Northeast Blackout." *The Atlantic*, 13 August 2018.

This triggered a chain reaction that led to power outages across the power grid. Like Per Bak's avalanche, a tiny grain of sand in the form of overgrown trees caused the carnage. That's pretty tough to predict.

Many people believe that earthquakes, forest fires, stock market crashes, the power grid, animal extinction, and political upheavals are like the sand pile game. And some people believe that the modern world creates more sand pile games relative to simpler times. Captain Smith of the RMS Titanic knew that his ship had the most modern technology built by expert engineers. He believed that the ship was "unsinkable." Did that make him less conscious of icebergs? Does wearing your seat belt make you safer or more likely to drive aggressively and thus create a sand pile game? What about the autopilot function on commercial planes? Today's pilots do not fly as much as their predecessors because of the ample use of autopilot functions. In those rare and unlikely cases when they are forced to fly without autopilot... will they have the Captain Sully skills? Will the flight simulators match up to real world problems? Have the "safety" precautions created a sand pile game? Questions like these are impossible to answer. But people should be cognizant of potential sand piles and wary of black swans. Identifying these cases is difficult; patterns that look meaningful may be simply random.

Streaks

The 2007 Super Bowl pitted the National Football Conference (NFC) champion Chicago Bears against the American Football Conference (AFC) Indianapolis Colts. All Super Bowls begin with a coin flip to decide who gets the ball first. You will hear raucous cheers from the stands when the coin flips because millions of people have proposition or "prop" bets on the outcome. A prop bet is a side bet where you place a bet on something that doesn't influence the outcome of the game like a coin flip or the length of time it takes to sing the national anthem. Tails landed four straight years from 2003 through 2006, but the coin landed on heads in 2007. The NFC's Chicago Bears won the flip. Announcer Jim Nantz said, "Well, the odds are more than 1,000 to 1 but that's now 10 straight Super Bowls that the NFC has won the coin toss." Did Jim get it right?

Jim sort of got it right... but not really. Four outcomes can occur when flipping a coin 2 times: 2 outcomes from the first flip times 2 outcomes from the second flip (2 × 2) giving you the four possibilities of HH, HT, TH, and TT. Only one of those options is all heads, so the probability is 25% that you will flip two heads in a row. When you flip a coin 10 times the logic stays the same, but the numbers get messier. So, the possible number of outcomes from flipping a coin 10 times is equal to $2 \times 2 \times 2 \times 2 \times 2 \times 2 \times 2 \times 2 \times 2 \times 2 = 2^{10} = 1,024$. Only one of those options is all heads, so Jim's analysis is correct in that respect. Ten straight heads have more than 1000 to 1 odds.

But Jim Nantz made his comment during Super Bowl XLI or 41 meaning that this string of coin flips had some chances to occur. The probability of having a streak of ten straight coin flips anywhere in a total of 41 coin flips is 1.6%, which is between a 1 in 50 and a 1 in 100 shot… more like picking the ace of spades from a deck of cards or hitting a number on a roulette wheel, not a black swan.

Let's compare this to height. Getting ten straight heads in a row is equivalent to you walking out on the street right now and the first person who walks by being taller than 6 feet, 6 inches. While getting ten heads in a row anywhere in 41 flips of the coin is the next person being taller than 6 feet, 3 inches. But that's not all. Suppose the AFC won the coin flip ten straight years, we can be pretty confident that Jim Nantz would have said the same "1,000 to 1" remark only replacing AFC with NFC. Or suppose the coin had come up with ten straight heads or ten straight tails, we can be confident that Jim throws the "1000 to 1" comment into the broadcast. But the odds of any of those four things happening in 41 flips of the coin is 1.6% + 1.6% + 1.6% + 1.6% = 6.4%. That's a little bit less than the likelihood of picking any ace out of a deck of cards or the next male you walk by being taller than 6 feet, 1 inch. That's not the sinking of the Titanic.

Streaks are more common than our brains are hard-wired to believe. And we like to find patterns in streaks even when they don't exist. The probability of a streak of 10 heads in 100 flips is 4.4%. Shaquille O'Neal's career 52.7% free throw percentage is among the worst in league history. So bad that the "Hack a Shaq" defense was invented where players strategically fouled Shaq to force him to take free throws. When Shaq entered the NBA in 1992, the most consecutive misses at the free throw line belonged to Chris Dudley. Dudley, a Yale graduate whose basketball skills consisted of being 6 feet, 11 inches and grabbing rebounds, missed 13 consecutive free throws to set the record. That's one record that Shaq did not want. If we assume that Shaq's free throws are equivalent to coin flips such that pure chance determines a make or miss and we use the 11,252 free throws that Shaq took in his career, then Shaq had a roughly 50–50 chance of breaking Dudley's consecutive missed free throw record. Shaq did not get the record, he got lucky.

Random events like free throws and coin flips have streaks and humans will almost always invent some causal reason why. When Shaq makes 7 or 8 free throws in a row, announcers are more than willing to declare that his form has improved, his injuries have healed, or he's fighting less with Kobe. When your team wins ten games in a row the coach becomes a genius, the beat-up hat you wore during the game becomes a harbinger of good fortune, and average players become hall of famers. Nobody wants to believe that chance and randomness led to the streak.

Cartesian Anxiety Revisited

So, how do you know the difference between a black swan versus random coin flips of ten heads in a row? How do you know if the sand pile game is lingering? How do you know what is random and what is a pattern? What system is like a normal distribution versus a power law versus a black swan versus the sand pile game versus a streak of coin flips? The painful answer is that statisticians don't know. The proverbial silver bullet does not exist. Nobody knows. Black swans and sand pile games and the normal distributions are metaphors. They are narratives about how the world works. They are models. Statisticians, like detectives, are always groping through darkness making best guesses with hazy evidence. Statisticians are guessing which metaphor fits the situation they're studying.

Humans love certainty, but statistics do not provide certainty. Science does not provide certainty. Cartesian anxiety is the human frustration with the reality that the complicated world we live in is uncertain, and science cannot help much. We cannot tell if the next flip will be heads or tails. The world is random, black swans abound. Uncertainty rules. And humans are poorly designed to deal with this uncertainty. Yale psychology professor Laurie Santos argues that our brains react to uncertainty with the "sympathetic nervous system." The same system that invokes fear when a tiger is lurking in the brush. She says, "Our minds hate uncertainty. Even if you give people two really good options, but they don't know which really good option they are going to get, our brains activate in ways as if though we're feeling afraid of something. We're literally activating fear regions of our brain, even for two good things."[6] So, if you are given a coin flip where heads you get $50,000 and tails you get a new car… two outstanding outcomes… your body will still break out the sympathetic nervous system to deal with the unknown and react as if a tiger is about to have you for lunch. We really hate uncertainty.

Statistics help us make intelligent guesses. Statistics do not solve uncertainty. We cannot predict the next black swan. We do not know what side the

---

[6] Boodhoo, Niala interview with Laurie Santos, The Happiness Lab, podcast audio, 6 November 2020.

coin will land on. We're as clueless as Cal Nathan Hockley mistreating Rose Dewitt Bukater before inviting her third class love interest to the first class dinner. We can, however, use statistics to make better decisions, to understand the world better. We can carry different narratives in our minds when analyzing real world problems. Is this ship unsinkable? Are those ten straight wins because my team was great, or just lucky? How do I invest for retirement? Was the 2008 financial crisis a black swan or were the increases in bank leverage driven by years of low interest rates more like a sand pile game? Statistics cannot answer these questions, just give you a better way of thinking about them. We need to be careful and open-minded and willing to believe multiple narratives. And that's not easy.

coin will land on. We're as clueless as Cal Nathan Hockley mistreating Rose Dewitt Bukater before inviting her third class love interest to the first class dinner. We can, however, use statistics to make better decisions, to understand the world better. We can carry different narratives in our minds when analyzing real world problems. Is this ship unsinkable? Are those ten straight wins because my team was great, or just lucky? How do I invest for retirement? Was the 2008 financial crisis a black swan or were the increases in bank leverage driven by years of low interest rates more like a sand pile game? Statistics cannot answer these questions, just give you a better way of thinking about them. We need to be careful and open-minded and willing to believe multiple narratives. And that's not easy.

5

False Positives

Kind of Pregnant

The 2007 movie *Juno* begins with 16-year-old Juno MacGuff taking a pregnancy test and declaring, "God, that little pink plus sign is so unholy." Explaining her conundrum to Rollo the store clerk, Juno says: "I think the first one was defective. The plus sign looks more like a division symbol, so I remain unconvinced." Juno wants the test to show a positive sign for pregnancy without her actually being pregnant. Juno needs a false positive. But pregnancy tests cannot be "defective," can they?

After Paulie Bleeker fertilizes Juno's egg, Juno's body changes. She produces a hormone called Human Chorionic Gonadotropin or HCG. Pregnancy tests assume that a woman is not pregnant, and look for evidence that a woman is pregnant. When the HCG in a women's urine exceeds a specified level, the plus sign pops up indicating pregnancy. The American Pregnancy Association assumes that the specified level is 25 HCG units per milliliter.[1] If the HCG in the urine says 26 or more, then the unholy plus sign appears. But many things can, and do, go wrong.

The results of Juno's pregnancy test can end in four outcomes: two correct and two wrong. Juno is pregnant, and the test says she's pregnant—correct. Juno is not pregnant, and the test says she's not pregnant—correct. And the two wrong outcomes: Juno is not pregnant, but the test says she's pregnant—the false positive. Or Juno is pregnant, but the test says she's not pregnant—the false negative.

---

[1] https://americcanpregnancy.org

© The Author(s), under exclusive license to Springer Nature Switzerland AG 2024
R. T. Stewart, *Adventures in Statistics*, Copernicus Books,
https://doi.org/10.1007/978-3-031-61284-8_5

	Juno is not pregnant	Juno is pregnant
Juno's test says she is not pregnant	Correct	*False Negative*
Juno's test says she is pregnant	*False Positive*	Correct

Fertility drugs and other medications can trigger a women's production of HCG. And some medical conditions including kidney disease and ovarian cancer can up HCG levels. In these cases, a high HCG reading may cause the plus sign indicating pregnancy even when there is no actual pregnancy—a false positive. Problems can occur in the opposite direction as well. If you shake the pregnancy stick too much or drink too much Sunny Delight, you may artificially reduce the HCG reading. The test may read low HCG levels, causing the negative sign to pop up even though you are pregnant—a false negative. Defective tests leading to false positives and false negatives happen for all kinds of reasons. Pregnancy sticks get exposed to heat or moisture or simply expire. Also, women do not produce equal amounts of HCG any more than women have the same height. A healthy pregnant woman can have a low level of HCG and a negative test—or vice versa. Tests can be wrong.

Juno wanted a false positive, she wanted the test to say she was pregnant when she was not pregnant. But false positives are not always nonthreatening. Suppose Juno went to the doctors not for a pregnancy test, but for a cancer test. Healthy patients have received false positives and been told that they had cancer despite not actually having cancer. People have even gone through chemotherapy and taken strong drugs with strong side-effects because of false positives… not to mention the stress and trauma involved in believing that you have a debilitating disease.

False negatives are even more daunting. If Juno is pregnant and the test comes back with a false negative such that she does not believe that she's pregnant when she really is pregnant, then she might imbibe at the local teenage keg party, smoke a pack of Marlboro Reds, and try some wacky tabacky. Worse, if she and her doctor do not believe that she has cancer or aids or tuberculosis or diabetes or heart disease because of a false negative, then, she will not get the treatments that could save her life.

Pregnancy tests are what statisticians call a hypothesis test where you run a "test" to solve a yes or no question—are you pregnant? In hypothesis tests, you must define the "null hypothesis" or the condition you will be looking to find evidence against. In the case of a pregnancy hypothesis test, the null hypothesis is that Juno is not pregnant, and you look for evidence in the form of high levels of HCG in Juno's urine that would lead you to believe that she is pregnant. Definitive proof does not exist in statistics. The chance of a false positive or a false negative is always there which means that scientists always face tradeoffs.

Pregnancy tests and all hypothesis tests and all science involve tradeoffs. We know that high HCG levels indicate a large likelihood of pregnancy, but what level is "high"? In the example above, we chose anything above 25 as high, but we could have chosen 20 or 32 or 9 or 35 or 17. The higher the number we choose, the less likely that we will have a false positive mistake, but the more likely that we will have a false negative mistake. The lower the number we choose, the less likely that we will have a false negative mistake, but the more likely that we will have a false positive mistake… a Catch-22.

Hypothesis tests are confusing. But the main concept to understand is that hypothesis tests involve tradeoffs. Lowering the likelihood of making one mistake (false positives) will increase the likelihood of making another mistake (false negatives), and vice versa. Science is about tradeoffs which is why science is so hard. The magic bullet reducing the false positives and the false negatives does not exist. Tradeoffs are reality, certainty an illusion. So take more than one pregnancy test and get second opinions at the doctor, tests can be defective.

Kind of Guilty

Nicole Brown Simpson and Ron Goldman were murdered on June 12, 1994. Nicole's ex-husband, O.J. Simpson was charged with murder. The trial had it all: celebrity, money, violence, sex, and racism. America was enthralled. On October 3rd 1995, the verdict came back 'not guilty' and O.J. walked out of Los Angeles County Superior Court a free man—stunning.

Science is not the only place we use hypothesis tests. Courtrooms must make yes or no decisions just like pregnancy tests. Courtrooms decide 'guilty' or 'not guilty.' Pregnancy tests assume the woman is not pregnant and we look for evidence in the form of high levels of HCG in urine. Courtrooms assume that the defendant is not guilty and we look for evidence in the form of testimony, DNA, historical patterns, blood droplets, cuts, and properly fitting gloves. Like pregnancy tests, court cases can have four outcomes, two correct and two wrong.

	O.J. is not guilty	O.J. is guilty
The jury determines that O.J. is not guilty	Correct	*False Negative*
The jury determines that O.J. is guilty	*False Positive*	Correct

Most judicial systems throughout the world are set up so that the individual accused of a crime is innocent until proven guilty. The null hypothesis is that the individual is innocent, and the preponderance of evidence must be beyond a reasonable doubt. Did the O.J. jury get it right? Or did the jury find that O.J. was not guilty when he really was guilty—a false negative.

The prosecution began the trial by hammering the jury with sordid accounts of O.J. beating and demeaning Nicole during their tumultuous relationship, including playing the 911 calls that you can listen to anytime on YouTube. The defense did not deny that O.J. abused Nicole. Rather, the defense parried by citing statistics. Defense attorney Alan Dershowitz wrote in his book: "We knew we could prove, if we had to, that an infinitesimal percentage—certainly less than 1 of 2500—of men who slap or beat their domestic partners go on to murder them."[2] The defense confused the jury using the prosecutor's fallacy.

The fallacy distracts the jury with a meaningless probability while ignoring the "true" probability the jury should be assessing. So, Dershowitz hammered the jury with the probability of a man killing his wife given that he had previously beaten his wife—a small number (1/2500 or much less than 1%). But this probability is meaningless. Nicole Brown Simpson was murdered. The jury needed to assess the probability that a man had killed his wife given that he had previously beaten his wife AND that his wife was murdered. The prosecutor's fallacy is easy to fall victim to, but the logic is as broken as going from the statement "if you're Tiger Woods, then you play golf" to "if you play golf, then you're Tiger Woods." Let's illustrate with an example.

Suppose we have 100 married women with a history of being abused by their husband. Now suppose that 10 of the 100 women are murdered. Dershowitz is arguing that the probability of a man killing his wife given that he previously beat his wife is less than 10%. This is true in this farfetched example. But this probability is irrelevant to the jury. The relevant probability only looks at the 10 women who were murdered. Nicole Brown Simpson was murdered. How many of those 10 women—who were both abused by their husbands and murdered—were killed by their husbands? Gerd Gigerenzer estimates that 9 out of 10 battered AND murdered women were killed by their husbands in his book *Calculated Risks*. He uses data from the Uniform Crime Report published by the FBI to make these estimates. This is the relevant probability, and it's a lot larger than "1 of 2500."[3]

---

[2] Dershowitz, Alan M. *Reasonable doubts: The criminal justice system and the OJ Simpson case*. Simon and Schuster, 1997.

[3] Gigerenzer, Gerd. *Calculated risks: How to know when numbers deceive you*. Simon and Schuster, 2015.

Prosecutors should have jumped all over Dershowitz's faulty statistics and poor reasoning. Prosecutors should have screamed to the high heavens that the probability of the husband killing the wife—given that the wife was abused AND was murdered—is exceptionally high. Prosecutors should have repeated ad nauseam that reasonable estimates land around 90%. Prosecutors failed. And O.J. Simpson walked. He may be the most famous false negative in history.

Kind of Confusing

Medical doctors in the United States must survive a labyrinth of tests, classes, and internships including the Medical College Admission Test (MCAT), 4 years of medical school and laboratory work, rotations at clinics and hospitals, residencies, and medical board-certified exams... before earning a medical license. Navigating this maze takes brains and diligence. Doctors are smart and hard working. But a doctor named David M. Eddy wrote a paper in 1982 describing the mistakes that doctors make with hypothesis testing.[4]

Dr. Eddy asked medical doctors statistical questions surrounding hypothesis testing related to breast cancer in female patients and found that the doctors failed to grasp some important concepts. "The evidence presented shows that physicians do not manage uncertainty very well, that many physicians make major errors in probabilistic reasoning, and that these errors threaten the quality of medical care," wrote Dr. Eddy. Ouch. Statistics is hard, even doctors screw it up.

A mammogram uses X-rays to scan breasts for deposits of calcium or microcalcifications that indicate cancer. Like Juno's pregnancy test, a mammogram can have four outcomes:

	Patient does not have cancer	Patient has cancer
Patient's test says she does not have cancer	Correct	*False Negative*
Patient's test says she has cancer	*False Positive*	Correct

Dr. Eddy asked physicians to answer a problem about hypothesis testing. The doctors received three pieces of information:

---

[4] Eddy, David M. (1982). Probabilistic reasoning in clinical medicine: Problems and Opportunities. In D. Kahneman, P. Slovic & A. Tversky (Eds.), *Judgment under uncertainty: Heuristics and biases* (pp. 249–267). Cambridge, England: Cambridge University Press.

1. The probability that a woman has breast cancer is 1%.
2. The probability that cancer is correctly detected by the test is 80%.
3. The probability that the test incorrectly detects cancer when the patient does not have cancer is 10%.

The doctors were then asked: If a woman tests positive, what is the probability that she has breast cancer? More than 90% of the physicians got the answer wrong.

The answer is not straightforward. The clearest way to answer the question is to avoid probabilities in favor of frequencies. So, pretend that you have tested 1000 women and break them down into meaningful groups. Humans think more clearly using frequencies because probabilities are confusing. Whenever possible, think of frequencies not percentages; Shaq made 9 of his 18 free throws, not Shaq shot 50% from the line. Sentence one tells you that 1% of women have cancer, so 10 women have cancer (1,000 × 0.01 = 10) and 990 are cancer free.

	Patient does not have cancer	Patient has cancer	Totals
Patient's test says she does not have cancer			
Patient's test says she has cancer			
Totals	990	10	

Of the 10 patients who have cancer, sentence two tells us that 80% or 8 of the 10 with the disease will be correctly determined by the test. The remaining two are the false negatives; they have cancer, but the test did not detect that they have cancer.

	Patient does not have cancer	Patient has cancer	Totals
Patient's test says she does not have cancer		2	
Patient's test says she has cancer		8	
Totals	990	10	

Sentence three gives the remaining information needed to fill in the table by giving you the false positive rate which equals 10%. Thus, of the 990 cancer free patients, 99 (999 × 0.10 = 99) will test positive for cancer despite being cancer free. The remaining 891 (990 - 99 = 891) make up the difference such that the chart is complete.

	Patient does not have cancer	Patient has cancer	Totals
Patient's test says she does not have cancer	891	2	893
Patient's test says she has cancer	99	8	107
Totals	990	10	

Now we have the information necessary to answer the question: If a woman tests positive, what is the probability that she has breast cancer? Only the "Patient's Test Says She Has cCancer" row in the chart is relevant.

	Patient does not have cancer	Patient has cancer	Totals
Patient's test says she does not have cancer	891	2	893
Patient's test says she has cancer	99	8	107
Totals	990	10	

Of the 107 patients who tested positive, 8 or 7.5% $\left(\dfrac{8}{107} \right)$ actually have cancer. This conditional probability shows that when the overall rates are small such as the 1% in this example and the false positive rates are high (10%), then, a positive test means that you still have low odds of having the disease. So, be careful, and get second opinions. Take multiple tests. Take different types of tests. As Juno learned, when multiple tests show the same outcome, you're less likely to be in the false positive or false negative buckets.

Hypothesis Testing

Hypothesis testing is science. The scientific method consists of developing a hypothesis, collecting data about the hypothesis, and analyzing the data. Then, the hypothesis is either rejected or kept to investigate further. Science can disprove hypotheses and lend support and evidence to theories. But science cannot prove anything. As Einstein observed:

> *The scientific theorist is not to be envied. For Nature, or more precisely experiment, is an inexorable and not very friendly judge of his work. It never says 'Yes' to a theory. In the most favorable cases it says 'Maybe,' and in the great majority of cases simply 'No.* [5]

---

[5] Gaither, Carl C. *Gaither's dictionary of scientific quotations*. New York: Springer, 2012.

Remember the black swan. No matter how many white swans you see, you cannot prove that there are no black swans. This implies that repeated tests and different types of tests are necessary for meaningful analysis. But taking tests comes with costs too, even pregnancy tests.

High levels of HCG indicate pregnancy, but high levels of HCG can also indicate cancer. False positives in pregnancy tests have led to patients needlessly undergoing chemotherapy and unnecessary surgeries including hysterectomies.[6] Doctors have found high levels of HCG in women who were not pregnant and concluded that this false positive result indicates a rare form of cancer called gestational trophoblastic tumor. This may be the case for some people, but 22-year-old Jennifer Rufer was a healthy woman who received a false positive pregnancy test. The doctors concluded that she had cancer, so she underwent chemotherapy and a hysterectomy. But she did not have cancer. She had high levels of HCG the same way that Shaquille O'Neal is taller than most people. She was healthy. She underwent chemotherapy and devastating surgery for no reason. Jennifer won a $16 million lawsuit against the hospital, but few people feel like Jennifer won anything.

Cancer screening may do more harm than good. If we increase screening tests, we may catch more cancer in early stages. But more screening leads to more false positives and "over-diagnosis" and "over-treatment."[7] Cancer screening can detect a slow-growing, non-life threatening form of cancer that will not cause any problems during the person's lifetime. But the screening causes treatment and the associated physical harms that come with treatments like chemotherapy or surgery or any drug intakes combined with the psychological anguish associated with a cancer diagnosis. "Over-diagnosis" and "over-treatment" cause real harm to patients. Screening tests should come with sober judgment of the benefits and harms associated with the tests and treatments. Screening tests involve tradeoffs. Life is complex.

A checkup includes all kinds of hypothesis tests including blood, urine, blood pressure, cholesterol levels, and electrocardiograms (EKGs). The more tests you take, the greater the chance of receiving a false positive result. If hypothesis tests are set up such that any single test creates a 5% chance of a false positive, then, ten tests give you a 40% chance of a false positive. The

---

[6] Cole, Laurence A., Kirsi M. Rinne, Shohreh Shahabi, and Aziza Omrani. "False-positive hCG assay results leading to unnecessary surgery and chemotherapy and needless occurrences of diabetes and coma." *Clinical chemistry* 45, no. 2 (1999): 313–314.

[7] Saquib, Nazmus, Juliann Saquib, and John PA Ioannidis. "Does screening for disease save lives in asymptomatic adults? Systematic review of meta-analyses and randomized trials." *International journal of epidemiology* 44, no. 1 (2015): 264–277.

Prasad, Vinay, Jeanne Lenzer, and David H. Newman. "Why cancer screening has never been show to 'save lives'—and what we can do about it." *BMJ: British Medical Journal (Online)* 352 (2016).

more tests you take, the more likely that the false positive occurs. False positives cause anxiety and unnecessary treatment. But the panacea does not exist. We should take tests to determine the presence of treatable diseases, but more tests lead to more false positives… more cases like Jennifer Rufer. The answer is not clear because tradeoffs are the rule.

Real life is more complicated than a two-by-two chart. In real life, the charts do not have all the buckets filled out. In real life, we guess at the false positive and false negative rates. In real life, cancer screening is fraught with uncertainty. In real life, we must make tough choices with unknown consequences. Real life means tradeoffs. We need hypothesis tests and medical screenings to treat diseases. But treating diseases must entail more than just hypothesis tests. Medical professionals should use multiple tests and multiple techniques to make decisions. Never believe that one number or one test is meaningful. Always ask for something to compare it with. Always explore problems from multiple angles. Science is hard.

6

Counting

Counting Cows

Police in England stopped Paul McCartney on June 14, 1963 for driving over the speed limit. Because Paul had two previous speeding convictions, Alderman W.O. Hanford reprimanded him: "It is time you were taught a lesson." The court fined Paul 25 British pounds and prohibited him from driving for a year. During his driving ban, Paul used a chauffeur to travel. On his way to John Lennon's house to work on Beatle's music, Paul asked the chauffeur how he was doing. "Oh, working hard, working eight days a week," the chauffeur replied.[1] The Beatles released the song "Eight Days a Week" in December 1964 in the United Kingdom. In March 1965, the song went to number one on the Billboard Hot 100 US charts. The chauffeur did not receive a writing credit.

The chauffeur and the Beatles counted wrong. You cannot have 8 days in a week. But before judging, understand that counting is remarkably difficult. We learn to count at a young age where we get fooled into believing that counting is simple. But your teachers lied to you. Counting is easy in mathematics class because you're not counting anything. A child can recite 1, 2, 3… forever. But that type of counting is completely theoretical. When you get past math class, you must count real things and that's when problems arise. This is best illustrated with a farcical example.

Suppose the Beatles are touring America and a farmer needs help counting cows. Each Beatle must go down to the field, count the cows, and report back

---

[1] Jones, Kevin. "Celebrity skinned? Not in 1963." *The Sydney Morning Herald*, April 19, 2010.

© The Author(s), under exclusive license to Springer Nature Switzerland AG 2024
R. T. Stewart, *Adventures in Statistics*, Copernicus Books,
https://doi.org/10.1007/978-3-031-61284-8_6

to the farmer. Paul goes first and reports that there are 5 cows. John follows Paul and reports that there are 5½ cows because one of the cows is pregnant. George takes his turn and reports back that there are 4 cows and 1 calf. Finally, Ringo reports: 3 cows, 1 calf, and 1 bull. The Fab Four counted 5, 5½, 4, and 3. Who is right? Counting is not simple.

Not all cows are cows. Cow, bull, heifer, steer, and calf all have specific meanings. A cow is a full-grown female that is at least a year old and has given birth. A bull is a male that has not been castrated, is capable of breeding, and is at least 2 years old. A heifer is a young female that has not had a calf. A steer is a bull who has been castrated. A calf can be either male or female but must be less than a year or two old. These terms are the American vernacular, but different countries have different lingo for different types of cows including intact, micky, bullocks, stag, springer, and rig. Counting cows is hard.

All statistics require some sort of counting. But the cows show that counting involves choices around definitions. Pure numbers only exist in mathematics class. Everywhere else, counting involves making arbitrary decisions. Paul's chauffeur meant to amuse with the 8 days a week quip, but he was also fishing for a bigger tip by exaggerating his work schedule. Who is doing the counting matters. Advocates will expand their definitions to garner attention to the cause.

Take homelessness as an example. If a person spends one night on the street, are they considered homeless? Or is it 10 nights? Or 30 nights? Should the single mother who got evicted so she lives in her mother's basement count as homeless? Or suppose a fire destroys a wealthy California neighborhood, do we count these people as homeless? A broad definition will increase the count, making the problem seem worse. So, advocates for homelessness will argue for broad definitions. Before you count something, you must define it. And the definition makes all the difference. Advocates for the homeless or for cancer research of for suicide prevention or for child abuse or for countless other causes will use broad definitions, leading to large counts.

No definitions are perfect. Definitions involve tradeoffs. But be aware of how the thing being counted is defined and confirm that the definition is reasonable. Otherwise, the counts are useless, and the statistics stemming from those counts are meaningless. Paul's chauffeur was not worried about the accuracy of his count, he wanted a tip. Whenever you read a number, you should ask yourself three questions: *Who counted this? How did they count this? Why did they count this?*

Measuring

How did they count this? Before you count something, you must measure it. NBA rosters are limited to 15 players during the regular season and the league has 30 teams. Thus, 450 athletes are in "the league." Competition for those 450 spots is ruthless. Height is a huge advantage in basketball because the rim is 10 feet off the ground. The greatest basketball players of all time have all been taller than 6 feet, 5 inches including Michael Jordan, LeBron James, Bill Russell, Kareem Abdul-Jabbar, Magic Johnson, Larry Bird, Wilt Chamberlain, and Shaquille O'Neill. In a long-standing basketball tradition, players at all levels have been exaggerating their heights.

The NBA brass cracked down on height measurements in 2019 with a rule change. Before the change, heights got measured with shoes on and rounded to the nearest inch. Now, everyone must get measured barefoot. Many players shrunk overnight. Two stars from the 2018 NBA champion Golden State Warriors lost 2 inches each. Both Draymond Green and Klay Thompson went from 6 feet, 7 inches to 6 feet, 5 inches. But more curious were the players who went in the other direction. Former Golden State champion and NBA most valuable player Kevin Durant got taller. His official height went from 6 feet, 9 inches up to 6 feet, 10 inches. Durant has been fudging his height in "basketball circles" for years to avoid being pigeonholed as a power forward or center.[2] Measuring anything will lead to strange anomalies, not hard facts. Measuring is often more difficult than counting.

Michael J. Fox gained mega-stardom in the 1980s as both a television star on *Family Ties* and movie star in the *Back to the Future* series. At age 29, he learned that he was suffering from Parkinson's disease. Parkinson's disease is a disorder of the central nervous system that leads to shaking, stiffness, and difficulty with walking, balance, and coordination. In his book *No Time Like the Future*, he compares his trouble keeping his balance from Parkinson's with the German theoretical physicist Werner Heisenberg's 'Uncertainty Principle.' The principle states that the position and velocity of an object cannot both be measured exactly, at the same time, even in theory. Michael J. Fox writes:

> *Heisenberg describes exactly the conundrum that we Parkinsonians face in dealing with gait. I cannot determine my position and my velocity at the same time. It's an unsolvable problem; people caution me to slow down as I shuffle-step-stumble forward, but they don't realize that it's an impossible request. I simply can't feel how fast I'm going. Moreover, my brain won't let my body stop until it finds a safe position—*

---

[2] Herring, Chris. "Why NBA Players Lie About Their Height." *The Wall Street Journal*, May 5, 2016.

and it can't find a safe position while I'm still moving. This relates directly to Heisenberg's Principle. Joseph Heller would also lay claim to it as a classic Catch-22: It's a dilemma from which there is no escape, because of mutually conflicting conditions.[3]

The underlying problem that Michael J. Fox, Werner Heisenberg, and Joseph Heller are describing is that measurement is not lifeless. The act of measuring something often changes the measurement. If you stick a thermometer in a pot of water, the thermometer changes the temperature of the water, at least slightly. You cannot find the temperature without the thermometer, but using the thermometer changes the temperature. The act of measuring can change your measurement in meaningful ways. Measuring is hard.

Counting Big Numbers

John, Paul, George, and Ringo flew to America on February 7, 1964. This kicked off the British invasion as Beatlemania took over America's radios. At the time, the Ed Sullivan Show was a staple of American television. Airing on Sunday nights from 8 to 9 PM, the popular show followed Sunday dinner for millions of Americans from 1948 to 1971. The Beatles agreed to play on the Ed Sullivan Show three straight Sundays in February 1964 for a $10,000 fee and their travel expenses. More than 50,000 people requested tickets for the first show on February 9th that only 728 lucky fans attended. But more than 74 million Americans watched on television, 40% of the US population.

Who counted the 74 million people? Did somebody walk from house to house and apartment to apartment counting the number of people watching the Ed Sullivan show? And who counted the US population? Whenever you see a number on the news or in a book or on a website or in a tweet, ask three questions. Who counted this? How did they count this? Why did they count this?

How many people are there in the United States? The accurate answer is nobody knows. But the US constitution mandates that censuses occur every 10 years. Article I, Section 2 of the US Constitution:

Representatives and direct Taxes shall be apportioned among the several States which may be included within this Union, according to their respective Numbers, which shall be determined by adding to the whole Number of free Persons, including those

---

[3] Fox, Michael J. *No Time Like the Future*, Flatiron Books, 2020.

bound to Service for a Term of Years, and excluding Indians not taxed, three fifths of all other Persons. The actual Enumeration shall be made within three Years after the first Meeting of the Congress of the United States, and within every subsequent Term of ten Years, in such Manner as they shall by Law direct.

The US census did not include "Indians" when counting people. And slaves counted as "three fifths" of a person. Those are tough facts to digest 230 years later. But counting means people defining something. These definitions are arbitrary. This does not mean that all numbers should be discounted, but that you must always ask: *Who counted this? How did they count this? Why did they count this?* The early census in America failed to count Indians and slaves because the people doing the counting did not want to give away political power. And racist views were normal.

The modern-day US Bureau of the Census counts people of all races. Citizens, non-citizen legal residents, non-citizen, long-term visitors, and undocumented immigrants get counted. The decision on whom to count is based on the concept of "usual residence" which comes from the Census Act of 1970 and means the place a person lives and sleeps most of the time. For most people, this is straightforward. But many groups of people have transient living arrangements including college students, homeless people, prisoners, undocumented workers, and military personnel. These people are difficult to count. College students often get counted twice because they often have two residences. While undocumented workers may not get counted at all because they avoid interacting with anyone from the government. The Census Bureau is aware of these problems and they adjust their counts, but the census remains a rough estimate. Nobody knows how many people live in the United States.

A.C. Nielsen reported that 75 million people watched the Beatles on the Ed Sullivan Show in 1964. They did not go to all the households in America and spy on the television sets. How did they count this? Nielsen used what they term Nielsen families, designed to be representative of certain geographical areas. Then, Nielsen had these families keep diaries of what they watched or Nielsen installed meters on their televisions to track what the families watched. Nielsen used a sample. Samples work well as long as they are representative, but representative samples are difficult to create. How representative was Nielsen's sample in 1964? Did they get The Beatles number correct? There's no way to tell. But there are countless examples of samples that were not representative leading to counts and accompanying statistics and accompanying predictions that were meaningless.

Unrepresentative Samples

Unlike Nielsen ratings, samples used to predict vote counts in elections have their day of reckoning. Poor sampling leading to failed election predictions are notorious. In 1948, polling predicted that the Republican New York Governor Thomas Dewey would soundly defeat the Democrat incumbent President Harry Truman. Truman won. But not before the Republican leaning Chicago Daily Tribune printed on the front page in bold with all capitals: **DEWEY DEFEATS TRUMAN.** A printers strike forced the paper to go to press early, so the editors trusted the polls and made an infamous mistake. The press handed a copy of the paper to Harry while on a train making his way back to Washington. Then, a lucky photographer took a picture with President Truman holding up the Chicago Daily Tribune headline and sporting the biggest grin we've ever seen on any president before or since. The photo is a staple in every American history book.

Nobody knows why the polling was wrong, but theories abound. The sample may have been unrepresentative because of overuse of phone surveys. About half the voting population had phones at the time, and people with phones tended to be richer and more likely to vote Republican. Another theory is that the high poll numbers induced Dewey to campaign passively, trying not to make any mistakes. Truman took the opposite path and attacked the campaign vigorously in the last few weeks before the election—after the polling, but before the vote. The exact reasons why the polling was unrepresentative of the voting public are not known. The world is complicated and simple polls will often fail.

The 2016 election results were also disastrous for the pollsters. Nate Silver gained attention in 2008 when his statistical model correctly predicted 49 of the 50 states in the presidential election. Nate's website, fivethirtyeight.com, became a crucible for pundits, political junkies, and curious citizens looking for the latest poll predictions. In 2016, Nate's model predicted that Hillary had a 71% chance of beating Donald. That's roughly the same as getting at least one head in two flips of the coin, or Shaq making at least one free throw in four attempts. Trump won, and Nate got dragged through the mud. *Current Affairs* published an article called: "Why You Should Never, Ever Listen to Nate Silver."[4]

Nate Silver's reaction was more sanguine; he pointed out that he predicted a 29% chance of a Trump win … much higher than most predictions. Again,

---

[4] Robinson, Nathan J. "Why You Should Never, Ever Listen to Nate Silver." *Current Affairs*, December 29, 2016.

nobody is certain why the samples were not representative. And again, theories abound. And again, phones played a part. Only now, everyone who votes has a cell phone, and landlines are going the way of the horse and buggy. So, polling over phones is unreliable. Also, Trump's anti-establishment messages like fake news may have caused Trump supporters to avoid pollsters. Or maybe people lied to the pollsters, saying that they would vote when they didn't or vice versa. The reason the polls failed in 2016 is the same as the reason the polls failed in 1948: the world is complicated and simple polls will sometimes fail. Not much has changed in 68 years. Expect the polls to fail spectacularly again at least once in the next 20 presidential elections. You can count on that.

Even in controlled studies, sampling is treacherous. Nobel Laureate James Heckman explained the problem to Russ Roberts on his podcast "Econtalk" with an example involving random assignment of AIDS (acquired immune deficiency syndrome) drugs.[5] In the late 1980s and early 1990s an AIDS diagnosis felt like a death sentence. Rock Hudson, Ryan White, Freddie Mercury, and Eazy-E all died tragically. Medical practitioners worked on developing drugs to address the problem, but that took time which is something AIDS patients did not have. Professor Heckman explains that a double-blind study was conducted to test some of these drugs such that the doctors and the patients had no idea who got the real drugs and who got the placebo. The two groups could then be compared to find if the drug had any effect. This is the textbook way to produce comparative samples, a perfect study. But the patients in the study did not want to risk getting the placebo, so they exchanged drugs with each other reasoning that half a dose was better than none. The perfect sample turned into disaster. The researchers did not know who had taken what. The sample was not representative of the patients who had taken drugs versus the patients who had not taken the drugs. Comparisons were impossible. Unrepresentative samples occur for all kinds of unpredictable reasons.

Coronavirus (Covid-19)

Covid-19 rocked the world in the spring of 2020. The new disease panicked everyone because of the unknowns. All kinds of questions surfaced, and the world demanded answers. The most pressing question being the number of deaths. The world wanted counts. Counting cows is hard but determining a

---

[5] Roberts, Russ. Interview with James Heckman. *Econtalk*. Podcast Audio. January 25, 2016.

cause of death is near impossible. Harvard trained medical doctor Gilbert G. Berdine said:

> *One definition would be deaths where Covid-19 was the primary cause. An example would be a patient who has a positive polymerase chain reaction (PCR) test for the virus responsible for Covid-19 and a clinical picture of hypoxemia, bilateral pulmonary infiltrates on imaging, no obvious other cause, such as influenza or congestive heart failure (CHF), and who dies from progressive acute respiratory distress syndrome (ARDS).*

> *Another definition would be deaths where Covid-19 was a contributing cause but not necessarily the primary disease. An example would be a patient with diabetes or end-stage renal disease who develops an upper respiratory infection URI, has a positive PCR test, never recovers from the URI, deteriorates over many weeks to months and eventually dies.*

> *In this case the underlying cause of death was the diabetes or end-stage renal disease weakening the host defenses and the Covid-19 was the precipitating cause of acute illness and eventual death. Another definition would be a patient who dies, has a positive PCR test, but the Covid-19 clearly had nothing to do with the death. An example would be a trauma victim who had no respiratory symptoms prior to trauma and coincidentally has a positive PCR test.*[6]

That's not all. Defining Covid-19 deaths is even more complicated. Hospitals—like Paul's chauffer and vertically challenged basketball players—face unclear incentives that may lead to undercounting or overcounting. Uncertainty is the rule when counting Covid-19 deaths.[7] To overcome these problems, statisticians have suggested comparing deaths in 2020 to deaths in previous years. But exploring these "excess deaths" is not easy. Can we be sure that differences are not random, like flipping a few heads in a row? Making sense of Covid-19 numbers will take time. Counting is hard.

---

[6] Berdine, Gilbert G. "Covid Misclassification: What Do the Data Suggest?" *American Institute for Economic Research*, November 30, 2020.

[7] Ioannidis, John PA. "Over-and under-estimation of COVID-19 deaths." *European journal of epidemiology* 36 (2021): 581–588.

7

Correlation

Golden Ticket

Moneyball is a 2011 movie about baseball and statistics. A movie about statistics that received six Academy Award nominations: best picture, best actor, best supporting actor, best adapted screenplay, best sound mixing, and best film editing. Hollywood legend Brad Pitt stars in the film as Billy Beane, the general manager of the Oakland Athletics, who transforms his team using stats. The movie is based on a 2003 book by Michael Lewis, *Moneyball: The Art of Winning an Unfair Game*. The book that changed all sports. The book that taught the sports world about correlation.

Baseball scouts get paid to predict how well players will perform. The caricature of the old-time scout in the movie is Matt Keough. He's a former major league pitcher with decades of baseball experience who married a Playboy playmate. Keough explains his doubt about a prospect's abilities, "Ugly girlfriend means no confidence." When challenged by another scout about the logic of correlating baseball success with a player's girlfriend, Keough doubles down: "I'm just saying his girlfriend is a 6 at best." Political correctness is not a forte among baseball scouts. Before *Moneyball*, baseball scouting consisted of eye tests, pop psychology, and rating girlfriends. But statistics changed everything. The caricature of the *'Moneyball'* scout in the movie is Peter Brand with his Ivy League degree and economics background. He says in an early scene, "It's about getting things down to one number. Using the stats the way we read them, we'll find values in players that no one else can see." Moneyball is born.

© The Author(s), under exclusive license to Springer Nature Switzerland AG 2024
R. T. Stewart, *Adventures in Statistics*, Copernicus Books,
https://doi.org/10.1007/978-3-031-61284-8_7

Correlations are relationships between variables. The correlation coefficient is a statistic, a single number, which measures the degree to which two variables move in relation to each other. Correlation coefficients fall between -1 and +1. Correlation coefficients closer to -1 indicate a negative correlation such that as one variable increases, the other variable decreases. And correlation coefficients closer to +1 indicate a positive correlation such that as one variable increases, the other variable also increases. Correlation coefficients near zero mean that the two variables are not related. For example, baseball teams that score a lot of runs will win a lot of games. Runs are positively correlated with wins. The 2019 baseball season ended with the following relationship between runs and wins.

Team	Wins	Runs
Houston Astros	107	920
Los Angeles Dodgers	106	886
New York Yankees	103	943
Minnesota Twins	101	939
Atlanta Braves	97	855
Oakland Athletics	97	845
Tampa Bay Rays	96	769
Washington Nationals	93	873
Cleveland Indians	93	769
St. Louis Cardinals	91	764
Milwaukee Brewers	89	769
New York Mets	86	791
Arizona Diamondbacks	85	813
Boston Red Sox	84	901
Chicago Cubs	84	814
Philadelphia Phillies	81	774
Texas Rangers	78	810
San Francisco Giants	77	678
Cincinnati Reds	75	701
Los Angeles Angels	72	769
Chicago White Sox	72	708
Colorado Rockies	71	835
San Diego Padres	70	682
Pittsburgh Pirates	69	758
Seattle Mariners	68	758
Toronto Blue Jays	67	726
Kansas City Royals	59	691
Miami Marlins	57	615
Baltimore Orioles	54	729
Detroit Tigers	47	582

correlations to match the story. But the relationship is spurious, called a *spurious correlation*. A spurious correlation refers to a statistically significant relationship that appears in the data, but only due to coincidence. There is no causal factor. The relationship is random, coincidental. Statistical correlation is not causation.

A simple two-step process was used to find the spurious relationship between murder rates and baseball wins. First, run correlation coefficients between baseball wins and as many variables as you can find, so you can choose the variable with the correlation coefficient farthest away from zero and most likely to have a small p-value. You must get results below the 0.05 rule. Second, manipulate the data slightly. In this case, removing St. Louis from the data set allows for the statistically significant relationship because St. Louis has a high murder rate and a high number of wins—the opposite of the final finding. The reason used for eliminating St. Louis from the analysis—more than 3 standard deviations from the average—is standard practice, taught in many data analysis classes. Finding spurious correlations is easy. You give me the data, and I'll verify any story you want to tell, boss. Juking the stats is easy.

Spurious correlations are more likely to occur in studies with small sample sizes and liberal definitions. The sample of baseball teams in this example is 30. Strange correlations can happen when looking at a single season worth of wins, but the relationship between murder rates and wins will not hold over time. When you increase the sample size, the spurious correlations disappear—fail to be *statistically significant*. Similarly, willingness to omit data as with St. Louis in this example is a red flag for spurious correlations. Definitions and the counts that come with them must be robust, and ignoring data almost always means that the counts are useless. Look for these clues when analyzing research. Also, look for minimal impacts. A correlation can be statistically significant, but that does not mean that the relationship has a meaningful impact.

Hacking data to "prove" a theory is common, and spurious correlations are powerful tools for misleading people. Stanford economist Thomas Sowell said, "One of the first things taught in introductory statistics textbooks is that orrelation is not causation. It is also one of the first things forgotten."[3] Do t forget the lesson. Spurious relationships abound in all types of analysis uch more important than exploring baseball wins. Spurious relationships e driven trifling laws.

ll, Thomas. *The vision of the anointed: Self-congratulations as a basis for social policy.* Hachette
19.

And we can view this relationship graphically.

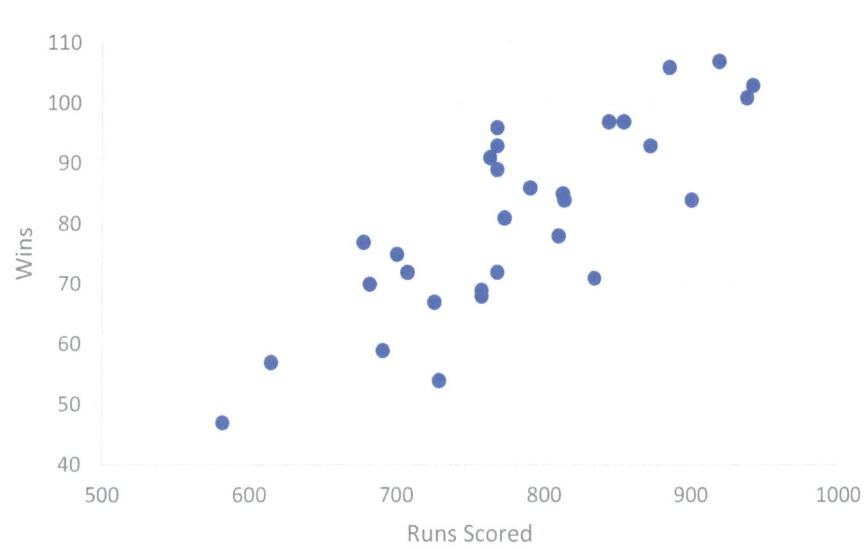

As runs increase, wins increase. The correlation coefficient is 0.82 meaning that there is a positive correlation between these two variables. Furthermore, the correlations coefficient is *statistically significant* because the p-value is less than the 0.05 rule. The 0.05 standard is a rule of thumb that statisticians have been using for close to 100 years. In short, statistical significance is jargon telling you that the relationship is unlikely due to chance. P-values less than 0.05 are necessary to be taken seriously in the statistical community.[1]

But the correlation coefficient is not perfect. Like the average family with 1.9 children, the 0.82 correlation is an abstraction that does not provide a complete story. The Houston Astros had the most wins but came in third in runs scored. The Boston Red Sox came in fourth in runs scored but fourteenth in wins. There are other variables that explain wins, most notably the number of runs allowed. The Red Sox offense was fine, but their defense was horrific. They needed better pitching. Also, correlation coefficients only capture linear relationships, so power laws and black swans are absent from the analysis.

Finding meaningful correlations is like turning tin into gold. If you find the golden ticket, the sky is the limit and statistics can make you a winner. Brad Pitt will play you in the movie. Baseball scouts are looking for qualities in players that correlate with future success. But even casual fans know who

[1] Gigerenzer, Gerd. "Mindless statistics." *The Journal of Socio-Economics* 33, no. 5 (2004): 587–606.

the best players are, so Billy Beane had to find correlations that were overlooked by other teams, and therefore, undervalued in the market. He needed inexpensive players. Peter Brand explains to Billy Beane that those players exist, "People are overlooked for a variety of biased reasons and perceived flaws: age, appearance, personality. Bill James and mathematics cuts straight through that. Billy, of the 20,000 knowable players for us to consider, I believe that there is a championship team of 25 people that we can afford, because everyone else in baseball undervalues them. Like an island of misfit toys." Brand references *Rudolph the Red-Nosed Reindeer* to make the argument that stats can find the correlations Billy needs, the correlations that other teams are failing to recognize. At the time, walks and on base percentage were undervalued in MLB. So, Billy hired a lot of no-name players who got a lot of walks.

With these misfit players, the A's won their division with 103 wins against 59 losses, and they broke the American League record for consecutive wins with 20 straight. The golden ticket. But the competition soon copied the A's success, and their advantage was "eroded within a year of *Moneyball's* publication."[2] But that's not the end of the story.

The sports world suddenly became statistical zealots. Teams began searching for hidden correlations to exploit. 'Moneyball' became a verb; the nerds took over. All sports have changed: we've "moneyballed" baseball, basketball, football, hockey, soccer, and rugby. Is "kick the can" next? Then, people argued for "moneyballing" education, healthcare, criminal justice, and the environment. It never seems to stop. 'Moneyball' is on fire. But finding meaningful correlations is not kids play.

Spurious

Runs are correlated with wins, and every baseball fan on the planet knows this. To acquire Billy Beane success, you must find correlations that are new and unknown. Table two shows the relationship between the murder rate per 100,000 people for 28 American baseball cities as recorded by the FBI in 2017 and the number of wins by baseball teams in those cities in 2018.

Team	Wins	Murder rate
Boston Red Sox	108	8
Houston Astros	103	12
New York Yankees	100	3

[2] Hakess, Jahn K., and Raymond D. Sauer. "An economic evaluation of the Moneyball hypothesis." *Journal of Economic Perspectives* 20, no. 3 (2006): 173–186.

Team	Wins	Murder rate
Oakland Athletics	97	16
Milwaukee Brewers	96	20
Chicago Cubs	95	24
Los Angelese Dodgers	92	7
Cleveland Indians	91	28
Colorado Rockies	91	8
Atlanta Braves	90	16
Tampa Bay Rays	90	10
Seattle Mariners	89	4
Pittsburgh Pirates	82	18
Washington Nationals	82	17
Arizona Diamondbacks	82	10
Philadelphia Phillies	80	20
Los Angeles Angels	80	7
Minnesota Twins	78	10
New York Mets	77	3
San Francisco Giants	73	6
Cincinnati Reds	67	23
Texas Rangers	67	12
San Diego Padres	66	2
Detroit Tigers	64	40
Miami Marlins	63	11
Chicago White Sox	62	24
Kansas City Royals	58	31
Baltimore Orioles	47	56

As the murder rate increases, the number of wins decreases. The correlation coefficient is negative 0.50 between the two variables. Furthermore, the correlation coefficient is *statistically significant* because the p-value is less than the 0.05 rule. Thus, the correlation is not due to chance. Cities with high murder rates tend to have poor baseball teams because star players do not want to l[ive] in those cities.

The city of Toronto was removed from the dataset because the FB[I] records cities in the United States. Similarly, the city of St. Louis was re[moved] from the data because the murder rate was more than three standa[rd devia]tions away from the average. Therefore, this data point is considered [an outlier] and removed from the analysis.

Stop. Stop. Stop. This analysis is garbage. No meaningful relat[ionship] between murder rates and baseball wins. That is nonsense. B[ut an ori]ginal statistician can make numbers tell stories and find corre[lations. Com]up with those stories. The story may be that cities with h[igh murder rates] cannot win baseball games because star athletes do not wa[nt to live in those] cities. Or the story may be that baseball players with un[...] lack confidence. The story can be anything. Rega[...]

[3] Sow[...]
UK, 2[...]

Abortion

Abortion ranks high among topics that spark emotions. Beliefs about abortion harbor personal experiences, religious doctrine, and political views. Dogmatists on both sides rile feelings to make their points. They also use— and abuse—statistics.

The 1973 Supreme Court decision in *Roe v. Wade* halted state laws that barred abortions to the chagrin of the state legislators who passed those laws. The states, therefore, fought *Roe v. Wade* by passing laws designed to make receiving an abortion more difficult. For example, the state of Texas passed the 'Women's Right to Know Act' in 2003. The law dictates that every abortion doctor warn the patient of "the possibility of increased risk of breast cancer following an induced abortion."[4] The law presumes a correlation between induced abortions and breast cancer, but that correlation may be spurious.

You can find research that shows correlations between induced abortions and breast cancer. The Texas Department of State Health Services (DSHS) references five papers.[5] DSHS uses this research to argue:

> *Your pregnancy history affects your chances of getting breast cancer. If you give birth to your baby, you are less likely to develop breast cancer in the future. Research indicates that having an abortion will not provide you this increased protection against breast cancer. In addition, doctors and scientists are actively studying the complex biology of breast cancer to understand whether abortion may affect the risk of breast cancer. If you have a family history of breast cancer or breast disease, ask your doctor how your pregnancy will affect your risk of breast cancer.*

And there is plenty of research arguing the opposite side. The American College of Obstetricians and Gynecologists (ACOG) references five papers to

---

[4] Health and Safety Code Chapter 171: Abortion. Texas Statutes.

[5] Huang, Y., Zhang, X., Li, W. et al. (2014). *A meta-analysis of the association between induced abortion and breast cancer risk among Chinese females*. Cancer Causes Control, 25: 227.

Jiang AR, Gao CM, Ding JH, et al. (2012). *Abortions and breast cancer risk in premenopausal and postmenopausal women in Jiangsu Province of China*. Asian Pac J Cancer Prev., 13:33–35.

Kamath R, et al. (2013). *A study on risk factors of breast cancer among patients attending the tertiary care hospital in Udupi district*. Indian J Community Med, 38(2)95–99.

Michels KB, Xue F, Colditz GA, Willett WC. (2007). *Induced and spontaneous abortion and incidence of breast cancer among young women: a prospective cohort study*. Archives of Internal Medicine; 167(8):814–820.

Reeves GK, Kan SW, Key T, et al. (2006). *Breast cancer risk in relation to abortion: results from the EPIC study*. International Journal of Cancer; 119(7):1741–1745.

argue that no correlations exist between induced abortion and breast cancer.[6] ACOG concludes:

> *Early studies of the relationship between prior induced abortion and breast cancer risk were methodologically flawed. More rigorous recent studies demonstrate no causal relationship between induced abortion and a subsequent increase in breast cancer risk.*

Do the women of Texas have the "right to know" anything about this research? Do political "leaders" have an obligation to provide a complete picture?

When research results contradict, scientists combine studies to find consistent and accurate results. In this case, the results of a holistic view of the research are overwhelming. The correlation between breast cancer and abortion is almost certainly spurious.[7] The American Cancer Society sums up the literature: "The issue of abortion generates passionate viewpoints in many people. Breast cancer is the most common cancer in women (aside from skin cancer), and it is the second leading cancer in women. Still, the public is not well-served by false alarms. At this time, the scientific evidence does not support the notion that abortion of any kind raises the risk of breast cancer or any other type of cancer."[8]

So codified laws exist based on meaningless statistics, spurious correlations. Big Brother decries that every woman in Texas (and other states with similar laws) who seeks an abortion has the "right to know" nonsense about a link between abortions and breast cancer. A correlation that does not exist. Be wary of numbers used to persuade rather than inform. Be wary of statistics that appeal to emotions. Be wary of statistics with skewed incentives. On June

---

[6] Rosenblatt KA, Gao DL, Ray RM, Rowland MR, Nelson ZC, Wernli KJ, et al. *Induced abortions and the risk of all cancers combined and site-specific cancers in Shanghai.* Cancer Causes Control 2006;17:1275–80.

Reeves GK, Kan SW, Key T, Tjonneland A, Olsen A, Overvad K, et al. *Breast cancer risk in relation to abortion: results from the EPIC study.* Int J Cancer 2006;119:1741–5.

Michels KB, Xue F, Colditz GA, Willett WC. *Induced and spontaneous abortion and inciNumber 434 June 2009 (Replaces No. 285, August 2003) 2 ACOG Committee Opinion No. 434 dence of breast cancer among young women: a prospective cohort study.* Arch Intern Med 2007;167:814–20.

Lash TL, Fink AK. *Null association between pregnancy termination and breast cancer in a registry-based study of parous women.* Int J Cancer 2004;110:443–8.

Henderson KD, Sullivan-Halley J, Reynolds P, Horn-Ross PL, Clarke CA, Chang ET, et al. *Incomplete pregnancy is not associated with breast cancer risk: the California Teachers Study.* Contraception 2008;77:391–6.

[7] Beral, V., D. Bull, R. Doll, R. Peto, G. Reeves, P. A. van den Brandt, and R. A. Goldbohm. "*Collaborative Group on Hormonal Factors in Breast cancer: Breast cancer and abortion: collaborative reanalysis of data from 53 epidemiological studies, including 83,000 women with breast cancer from 16 countries.*" Lancet 363, no. 9414 (2004): 1007–1016.

[8] ACS—American Cancer Society (Accessed 2020, December 8). *Abortion and Breast Cancer* [Online.] https://www.cancer.org/cancer/cancer-causes/medical-treatments/abortion-and-breast-cancer-risk.html

24, 2022, in a landmark case, *Dobbs v. Jackson Women's Health Organization*, and led by Judge Alito who wrote "an abuse of judicial authority" and "egregiously wrong" in the ruling, the Supreme Court overturned *Roe v. Wade*. Let's hope the lousy statistics did not play a part in the ruling.

Randomization

Ronald A. Fisher was born to a wealthy family in 1890 London, England. His father, George Fisher, was an art dealer who ran an auction company comparable to Sotheby's or Christie's. Ronald lived a life of plenty including attending the best schools in England until he was 14 in 1904. That year, Ronald's mother died of acute peritonitis. Then, his father's business failed. The family moved from a mansion in Hampstead to an average house in Streatham. But Ronald was able to continue his education when he won a mathematics competition and received a scholarship worth 80 pounds sterling. He eventually matriculated at Cambridge where he graduated with first class honors in mathematics.

R.A. Fisher suffered from poor eyesight—shortsightedness—his entire life. The disability drove how he digested mathematics. He could not learn like most students, and solved problems in his head rather than on paper. Strangely, his disability helped him. He mastered mathematics, and his professors considered him a genius. Later, math practitioners at the highest level recognized his talents. But R.A. Fisher was a lousy teacher. He struggled translating what seemed obvious to him. Thus, he took a job in 1919 with the Rothamsted Experimental Station in central England working on agricultural research.

The station ran experiments with different types of fertilizers for different crops. Typically, the farmers spread mixtures of phosphate and nitrogen salts over different fields, planted the crops, and measured the harvest. The next year, the farmers received instructions on changing the fertilizer based on the previous year's result and repeated the process. Researchers collected data on rainfall, temperature, fertilizers, soil quality, and harvests for over 90 years as this process played out. Fisher's employer wanted him to determine what mixtures of fertilizers across the different fields would produce the best harvests. Fisher could not complete the task.

The problem faced by Fisher when exploring the data was what he called "confounding." He could not tell if the size differences of the harvests stemmed from the differences in rainfall or the differences in fertilizers. So, despite almost a century of data, the effectiveness of the fertilizers could not be determined. The data showed an apples to oranges comparison. The data was useless. This led Fisher to think about how to design experiments and the concept of randomization.

Before Fisher, farmers fertilized entire fields and researchers examined year-over-year differences in harvests. This did not work because temperature and rainfall—confounding factors—could drive the difference rather than the fertilizer. There was no way to tell which factor was causing the quality of the harvests—the fertilizer or the rainfall or the temperature or the field or some unknown factor. Fisher argued that the fertilizer should be applied randomly across the same field. So, the farmers separated fields into subplots and fertilized rows within the subplots differently. Therefore, the temperatures, rainfall, and soil quality were "controlled for" in the experiment. We call this a *randomized control study* today, and it seems obvious. But it was not obvious for the 90 years of experiments that preceded R.A. Fisher. Nobody had thought of it.

Randomized control studies are the gold standard for scientific analysis. There are innumerable factors that drive harvests. Not just rainfall and temperature, but harvests are influenced by how the water drains, soil quality, weeds, animals, and thousands more factors. A researcher can never "control for" all factors. Randomization solves this problem. You do not have to know all the confounding factors. As long as the fertilizer gets applied randomly throughout the field, the confounding factors will cancel each other out. You will have an apples to apples comparison. You can make an argument for causation—the gold standard.

Smoking and Cancer

Ronald A. Fisher loved smoking. "And to take the poor chap's cigarettes away from him would be rather like taking away his white stick from a blind man. It would make an already unhappy person a little more unhappy than he need be," he wrote in 1958.[9] Fisher enjoyed the lost art of smoking a pipe, and he did so from a young age until he died in 1962.

While Fisher was enjoying his pipe, the British Medical Research Council became alarmed at the increase in the number of people dying from lung cancer. Improved sanitation and medical care meant longer lives and a healthier population as the twentieth century progressed, but lung cancer was moving in the opposite direction. More and more people across the globe were

---

[9] Fisher, Ronald. "Cigarettes, cancer, and statistics." *The Centennial Review of Arts & Science 2* (1958): 151–166.

dying from lung cancer. The British Medical Research Council hired Austin Bradford Hill and Richard Doll to find out why.

Hill and Doll hypothesized that smoking causes lung cancer, but they also suspected carcinogens from the tar in roads and the exhaust from cars. There were plenty of theories about why lung cancer seemed to be getting worse. So, Hill and Doll ran into the same problem that perplexed Fisher in the fields: confounding factors. They needed randomization. Randomly giving fertilizers to different plants in the same field allowed for fair comparisons, the gold standard. But you cannot sample 1000 people and randomly tell half to refrain from smoking while forcing the other half to smoke 30 cigarettes per day. How do you get an apples to apples comparison without randomization?

Because randomization was not possible, Hill and Doll used 'retrospective' and 'prospective' studies. A retrospective study finds a sample of people with the disease, then looks for conditions across patients that may be associated with the disease. Hill and Doll interviewed 1400 hospital patients in London hospitals, about half who suffered from lung cancer. They asked all kinds of questions and a pattern soon emerged: smoking and lung cancer seemed to be linked.[10] Midway through the interview process as the evidence became clearer, Richard Doll quit smoking.

Hill and Doll followed the retrospective study with a prospective study. They surveyed 30,000 medical doctors across England asking them about their smoking habits and medical histories. Then, they followed this group for 5 years. Again, the results showed that smokers were more likely to suffer from lung cancer than non-smokers.[11] The two papers had an impact and other researchers followed with similar studies. All seemed to find a link between smoking and cancer.

But R.A. Fisher was unconvinced. He continued smoking his pipe. Furthermore, he wrote and lectured about his unwillingness to believe that smoking causes cancer.[12] Fisher's primary argument was that the failure to use a randomized study opened the door for a confounding factor. For example,

---

[10] Doll, Richard, and A. Bradford Hill. "Smoking and carcinoma of the lung." *British Medical Journal 2*, no. 4682 (1950): 739.

[11] Doll, Richard, and A. Bradford Hill. "The mortality of doctors in relation to their smoking habits." *British Medical Journal 1*, no. 4877 (1954): 1451.

[12] Fisher, Ronald. "Cigarettes, cancer, and statistics." *The Centennial Review of Arts & Science 2* (1958): 151–166.

Fisher, Ronald A. "Lung cancer and cigarettes?" *Nature* 182, no. 4628 (1958): 108–108.

Fisher, R.A. "Cancer and smoking." *Nature* 182, no. 4635 (1958): 596–596.

Fisher, Sir Ronald Aylmer. *Smoking: the cancer controversy: some attempts to assess the evidence.* Edinburgh: Oliver & Boyd, 1959.

suppose that there is a genetic predisposition to smoke like a genetic predisposition for having red hair. And suppose that this genetic predisposition also increases your risk of lung cancer like red heads are more likely to suffer from skin cancer. In this case, the genetic predisposition—not the smoking—is causing the lung cancer. Fisher's main point seems to be that proving causation outside of randomized studies is challenging.

A single study—no matter how well designed—cannot prove anything. Every study has limitations, and someone with Fisher's intellect can legitimately question any study. Retrospective and prospective studies are not beyond reproach; they are not as convincing as a randomized control study. Statistics provide evidence, not proof. Like detectives, statisticians must look at all the evidence to draw conclusions. Smoking guns do not exist. In 1959, researchers from The National Cancer Society, the American Cancer Society, and the Sloan-Kettering Institute wrote a 30-page paper reviewing all the available studies including the work from Doll, Hill, and Fisher. The paper determined a link between lung cancer and cigarettes, the "evidence supports the conclusion of a causal relationship with cigarette smoking."[13] The medical community was convinced. No reputable journals have since published anything arguing that smoking does not cause lung cancer.

Fisher got it wrong. Fisher is the father of modern statistics. Fisher is a mathematical wunderkind. And Fisher got it wrong. Why? Researchers have pondered this question and come up with theories that Fisher was motivated by a combination of greed, arrogance, and a love of smoking.[14] But it's probably not that complicated. Fisher had decided that smoking was not harmful at a young age, and no evidence was going to change his mind. Fisher suffered from confirmation bias. We all suffer from confirmation bias.

René Magritte painted a famous picture called *The Treachery of Images* in 1929. The painting is of a generic pipe with the words: "*Cici n'est pas une pipe.*" The words translate: "This is not a pipe." The painting is a warning. The image is not the thing. Magritte said, "So if I had written on my picture 'This is a pipe,' I'd have been lying." It is paint on canvas. It is just a representation. The map is not the territory. The model is not the real world. The number is not the explanation. This is not a pipe. Humans cannot stop creating representations, categories, and explanations to make their worlds robust, less uncertain. But when we fall in love with our representations of the world, we

[13] Cornfield, Jerome, William Haenszel, E. Cuyler Hammond, Abraham M. Lilienfeld, Michael B. Shimkin, and Ernst L. Wynder. "Smoking and lung cancer: recent evidence and a discussion of some questions." *Journal of the National Cancer Institute 22*, no. 1 (1959): 173–203.
[14] Stolley, Paul D. "When genius errs: RA Fisher and the lung cancer controversy." *American Journal of Epidemiology* 133, no. 5 (1991): 416–425.

fall into the trap of confirmation bias. We select information that fits our narrative and ignore conflicting evidence. We are willing to believe that smoking comes without risks, or the attractiveness of the ballplayer's girlfriend matters, or that murders affect baseball wins. We all do this, even Ronald Aylmer Fisher. Science is not a straight line. Science is muddy and messy and ugly. And science is entirely human. Always remember that numbers are representations, not reality. This is not a pipe.

8

Juking the Stats

Good Luck

Liam Neeson plays former CIA officer Bryan Mills in the 2009 box office hit *Taken*. Bryan has a challenging relationship with his teenage daughter, Kim, because of his divorce and his obsession with work. Then, Kim gets kidnapped. During the kidnapping, Bryan speaks to the criminals on Kim's phone:

> *I don't know who you are. I don't know what you want. If you are looking for ransom, I can tell you I don't have money, but what I do have are a very particular set of skills. Skills I have acquired over a very long career. Skills that make me a nightmare for people like you. If you let my daughter go now that'll be the end of it. I will not look for you, I will not pursue you, but if you don't, I will look for you, I will find you, and I will kill you.*

"Good luck," replies the kidnapper. Viewers immediately know what the next 80 min will look like. Good guy, Bryan Mills, will find and torture the bad guys. The "good luck" guy will pay for that comment. Bryan and Kim will reunite with father and daughter bonds restored. The good guys win, and the bad guys get their comeuppance. Buckle up, this is going to be fantastic. How can you not love Hollywood?

David Simon created *The Wire*. But his background in journalism made him a failure at Hollywood scripting. He must have missed *Taken*. Audiences want the good guys to beat the bad guys, resolve their interpersonal relationships, and live happily ever after… all within 2 hours. David Simon did not get the memo. He must have missed *Die Hard*, *Lethal Weapon*, and *Fargo* too.

R. T. Stewart, *Adventures in Statistics*, Copernicus Books,
https://doi.org/10.1007/978-3-031-61284-8_8

These are cop stories worth watching. The good guys defeat the bad guys in less than 2 hours. Poor David did not have the particular set of skills to write those scripts. He had another plan.

The cop genre changed in 2002 when HBO premiered *The Wire*. This is not another cop show, but "a novel for television."[1] The characters are complex, the stories detailed, and the endings tragic. *The Wire* forces viewers to confront the reality that the good guys are not all good and the bad guys are not all bad. The line between good and evil stops functioning in the first episode and never recovers. The characters are nuanced, none more so than Roland Pryzbylewski who usually goes by "Prezbo" or just "Prez." Prez symbolizes *The Wire's* overarching mantra that people are complicated. Prez is simultaneously the least and most sympathetic character on the show.

For reasons that defy logic, Prez shoots his own police car, pistol whips a young boy in an act of police brutality, and fires his weapon inside a police office. Worse, he kills a fellow officer when he mistakes the officer for a perpetrator. And even Prez himself expresses his uncertainty around whether race influenced his poor decision making in the shooting. But Prez earns viewers sympathy when he becomes a middle school math teacher. He develops into a serviceable teacher who helps his students beyond the classroom with food and laundry. He even pushes the administration to improve the bureaucracy by demanding new books. Prez is complicated. Audiences do not know if they should hate him or love him.

Statistics are complicated too. Politicians, bosses, and the public demand the statistic that solves the problem, clarifies the argument, and makes them right. We want *Taken*. Exactly 90 minutes of awesome where the good guys win, the bad guys lose, and everyone goes home happy. We get *The Wire*. More than 60 hours of heartbreak where nobody wins, the bureaucracy carries on, and everyone goes home wondering why we cannot do better.

Goodhart's Law

Prez started as a police officer and became a middle school teacher. The constant across both institutions: "juking the stats." Juking the stats describes the practice of alternating and gaming statistics to make yourself or your boss look better. Prez explains his experience juking stats as a cop to his new

---

[1] Lynskey, Dorian, "The Wire, 10 years on: 'We tore the cover off a city and showed the American dream was dead.' *The Guardian*, March 6, 2018.

teaching buddy: "Making robberies into larcenies. Making rapes disappear. You juke the stats, and majors become colonels. I've been here before."

Economists were onto stats juking long before *The Wire*, but they called it Goodhart's Law.[2] Goodhart's Law states:

When a measure becomes a target, it ceases to be a good measure.[3]

A variant of Goodhart's Law is the 'cobra effect.' The cobra effect describes an anecdote about venomous cobras in Delhi, India during British colonial rule. The British governor was nervous about the large number of cobras in the city, so he declared a bounty for each dead cobra. Residents responded by bringing in dead cobras. The cobra population decreased. But enterprising residents went a step further; they began breeding cobras which were then killed to receive the bounty. When the governor learned of this scheme, he scrapped the bounty. But the entrepreneurs had cages filled with now worthless cobras. They solved this problem by releasing the cobras onto the streets. The cobra population skyrocketed.

The same thing happens in police departments across the country. As depicted in *The Wire*, CompStat is a management tool used in police departments that stands for compare statistics. The idea is that management can identify underperforming precincts using statistics and address crime spikes using targeted enforcement. But crime, like cobras in Delhi, had other ideas. CompStat started in New York City and middle level police commanders were soon under tremendous pressure to lower the crime statistics. They figured out that "underreporting" was an option. So, the police lied; they found ways to stop reporting crimes. "You could refuse to take crime reports from victims, you could write down different things than what had actually happened. You could literally just throw paperwork away," reported former NYC cops.[4] The crime statistics became garbage. Goodhart's law in action.

The same thing happens in schools across the country. "Teaching to the test" is a theme in season four of *The Wire*. Prez gets tough lessons from his students, but tougher lessons from the administration. Getting better test scores is the job, teaching the students is marginalized. As usual, *The Wire* mirrors real life.

---

[2] Goodhart, Charles AE. "Problems of monetary management: the UK experience." In *Monetary Theory and Practice*, pp. 91–121. Palgrave, London, 1984.

[3] Strathern, Marilyn. "'Improving ratings': audit in the British University system." *European review* 5, no. 3 (1997): 305–321.

[4] Vogt, PJ. "#128 The Crime Machine, Part II." *Reply All*. Podcast Audio. October 12, 2018.

George W. Bush signed the *No Child Left Behind Act of 2001* which promised to improve "the academic achievement of the disadvantaged." Like toddlers writing letters to Santa Claus, politicians ignored the complexities of education and presumed that a few words on a piece of paper in Washington would do the trick. The law expanded the role of the federal government in education and demanded more testing. Test scores would determine the amount of federal funds received. Politicians had the hubris and conceit to believe that they could "fix" education. But they failed to acknowledge Goodhart's law.

After *No Child Left Behind* passed, schools across the country doubled down on improving the testing stats. The pressure was on. And Goodhart's law did not disappoint. The teachers did the same thing the police did, the same thing the cobra entrepreneurs in India did. They cheated. You can cheat by giving students answers to the test before the test, providing answers during the test, and changing the answers after the test. You can also cheat by "scrubbing." Scrubbing occurs when marginal students are added or deleted from the rolls to make the testing scores look better. You get caught cheating when test scores improve dramatically or "erasure" analysis finds evidence of changing answers from wrong to right.[5]

Cheating scandals shook public education across the country. The FBI investigated. Racketeering charges usually reserved for mobsters and murderers got used against kindergarten teachers. At the behest of the FBI, teachers wore wires to trap other teachers. Teachers got arrested, tried, convicted, and sent to prison. The failure of *No Child Left Behind* was predictable. Many practitioners with knowledge of Goodhart's law anticipated the dangers in articles published before *No Child Left Behind* passed.[6] But the politicians got their inept law passed anyway.

We all want simple solutions to difficult problems. Pay for dead cobras; track arrest records; monitor school progress through testing. High level, top-down solutions are enticing. But Nobel laureate F.A. Hayek debunks our ability to find these solutions by highlighting the "*fatal conceit* that man is able to shape the world around him according to his wishes."[7] Top-down decision-making fails. The politicians trying to improve the police and the teachers have good intentions, but they fall short. Confident folks in Washington cannot solve problems with policing and education at the 50,000-foot level. The

---

[5] Jacob, Brian A., and Steven D. Levitt. "Rotten apples: An investigation of the prevalence and predictors of teacher cheating." *The Quarterly Journal of Economics* 118, no. 3 (2003): 843–877.

[6] Koretz, Daniel. *The testing charade: Pretending to make schools better.* University of Chicago Press, 2017.

[7] Von Hayek, Friedrich. *The fatal conceit*, p. 34. University of Chicago Press, 1988.

problems and the solutions are locally varied and complex. They can only be understood close up, on a case-by-case basis. The details matter. As Hayek wrote, "The curious task of economics is to demonstrate to men how little they really know about what they imagine they can design."[8]

Taken peaks when Bryan Mills finds his daughter held at knifepoint by the evil sheikh Raman. Bryan kills Raman with a single bullet to his forehead. Then, Bryan's daughter embraces Bryan with a giant hug. The good guys win; the bad guys lose—the perfect movie. We all want simple solutions to difficult problems.

Multiple Numbers

Hans Rosling wrote the book *Factfulness: Ten Reasons We're Wrong About the World—and Why Things Are Better Than You Think*. The book came out in 2018 to positive reviews from across the globe. "One of the most important books I've ever read; an indispensable guide to thinking clearly about the world," raved Bill Gates.[9] The book shatters misconceptions about the world using statistics and facts. Hans also has YouTube videos—with millions of views—explaining the statistics he uses to explain healthcare around the world. You do not get millions of views on YouTube for a video about statistics without an excellent product. He has it. His enthusiasm oozes off the pages in his book and in his talks. The statistics world has never had a better teacher.

Hans taught many statistical lessons in his work, but perhaps his most important lesson is also his simplest. He taught everyone to always look at more than one number. He commands:

The most important thing you can do to avoid misjudging something's importance is to avoid lonely numbers. Never, ever leave a number all by itself. Never believe that one number on its own can be meaningful. If you are offered one number, always ask for at least one more. Something to compare it with.[10]

He uses "never." There is no doubt. A number by itself, a lonely number, is useless. Never believe one number. Never.

---

[8] Von Hayek, Friedrich. *The fatal conceit*, p. 82. University of Chicago Press, 1988.

[9] Brueck, Hilary. 'The author of one of Bill Gates favorite books says the world isn't as apocalyptic as you might think—here are five of his top reasons why,' *Business Insider*, April 5, 2018.

[10] Rosling, H., Rosling, O., & Ronnlund, A. R. (2018). Factfulness: ten reasons we're wrong about the world—and why things are better than you think. First edition. New York. Flatiron Books.

Looking at multiple numbers allows us to avoid Goodhart's Law. If you do not want to be duped by data, always look at multiple numbers. Teaching to the test forces teachers, students, and administrators to concentrate on one number. But student abilities and a teacher's worth are too complicated for a single number. Never judge anyone with a single number. Baseball players are judged with multiple numbers: batting average, home runs, speed, defensive ability, on base percentage, and slugging percentage… to name a few. Baseball is a simple game. You throw the ball, you catch the ball, you hit the ball. But scouts now use dozens of numbers to judge baseball players. And they use ample qualitative analysis as well. Teaching is vastly more complicated than baseball. We cannot judge teachers with a single number. We cannot judge policemen with a single number. "Never." Always use multiple numbers.

Banners and Banquets

In Prez's rookie year as a middle school teacher, he takes Duquan "Dukie" Weems under his wing. Dukie's parents are drug addicts who fail to provide Dukie with food or clean clothing. Prez makes sure that Dukie gets lunches, laundry, and a place to shower in the school. Prez earns good karma after some pretty horrendous behavior, leaving viewers with cognitive dissonance. Then, Dukie comes back to visit Mr. Prezbo at least a year after graduating middle school. Dukie hustles Mr. Prez for money using lies about his desire to get back to school when his real intention is to buy heroin. Prez gives him the money even though he knows that Dukie is playing him, leaving viewers with tears in their eyes. Seems unlikely that any school could have improved Dukie's fate.

So, how do you improve schools? Nobody knows the answer, but you can start by getting the statistics correct. The Gates Foundation failed as Howard Wainer details in his book *Picturing the Uncertain World*:

> *Then in the late 1990s the Bill and Melinda Gates Foundation began supporting small schools on a broad-ranging, intensive, national basis. By 2001, the Foundation had given grants to education projects totaling approximately $1.7 billion. They have since been joined in support for smaller schools by the Annenberg Foundation, the Carnegie Corporation, the Center for Collaborative Education, the Center for School Change, Harvard's Change Leadership Group, the Open Society Institute, Pew Charitable Trusts and the U.S. Department of Education's Smaller Learning Communities Program. The availability of such large amounts of money to implement a smaller-schools policy yielded a concomitant increase in the pressure to do so, with programs to splinter large schools into smaller ones being proposed and imple-*

mented broadly (New York City, Los Angeles, Chicago and Seattle are just some examples).[11]

Nearly $2 billion and incredible brain power: Gates, Carnegie, and Harvard. Yet they missed the basics of sample size and supported smaller schools based on a misreading of data.

Nobody knows for sure why The Gates Foundation were adamant about smaller schools, but economists have speculated that they failed to understand the nuances around sample size. For example, many states have programs designed to highlight and reward schools that improve their test scores. In North Carolina, improved scores get you a "Top 25" reward and your school receives a banner and your teachers are given a banquet.[12] The following chart displays the percentage of the 840 elementary schools that received an award between 1997 and 2000 broken down by size. The bottom decile represents the smallest 10% or 84 schools, the second decile the next smallest 84 schools and so forth.

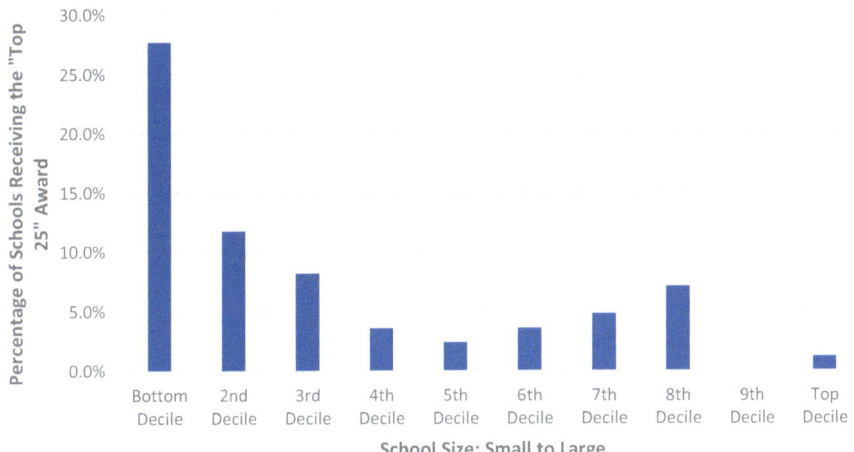

The results are overwhelming. The bottom decile—the smallest 84 schools in the sample—improved their test scores to earn the "Top 25" award 27.7% of the time while the largest decile sits at a mundane 1.2%. Small schools are great. We need smaller schools, smaller classrooms. Send the boatloads of money, Bill. Give the small schools banners and banquets.

---

[11] Wainer, Howard. *Picturing the uncertain world: How to understand, communicate, and control uncertainty through graphical display.* Princeton University Press, 2009.

[12] Kane, Thomas J., and Douglas O. Staiger. "The promise and pitfalls of using imprecise school accountability measures." *Journal of Economic Perspectives* 16, no. 4 (2002): 91–114.

Slow down, the logic is flawed. Small schools have greater variance in test scores. The spread or range or standard deviation around any statistic—including average test scores—changes depending on the sample size. To understand the way that sample size can confuse, imagine a strange world where each student is asked to flip a coin. Then, the teachers get rewarded for the number of heads the students receive. In this world, teaching talent does not exist, everything is determined by luck.

Now suppose you have a class of ten students, and all the students are asked to flip a coin. Using the computer, we can run this experiment 100,000 times. So, 100,000 imaginary classes of 10 students flip a coin and the heads are added up for each class. The teachers with the greatest number of heads get the banners and the banquets. With a 50% probability of flipping a head, there will be variance. You will have times that 3 heads or 6 heads or even 10 heads land in 10 flips of the coin. By letting the computer do the flipping for us, the percentage of heads in 10 flips of the coin over 100,000 trials will look as follows:

Number of heads	Percentage of heads (%)	Number of classrooms
0	0	97
1	10	1008
2	20	4453
3	20	11,634
4	40	20,344
5	50	24,534
6	60	20,726
7	70	11,679
8	80	4452
9	90	974
10	100	99

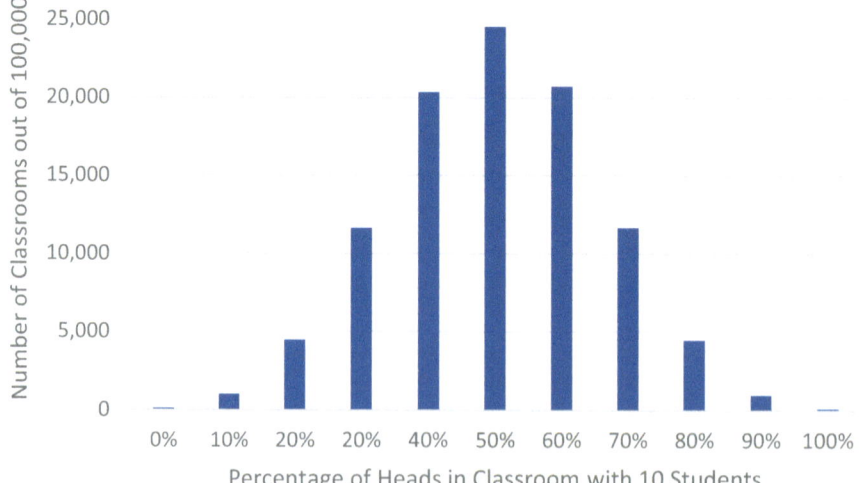

Percentage of Heads in Classroom with 10 Students

Heads have odds of 50%, but only 24,534 of our 100,000 trials ended up with 50% heads. There is variance, randomness, uncertainty… luck. So much so that 0.1% of the trials or 99 out of 100,000 in our sample ended up with 100% heads. Those schools will get banners and banquets. Still, 66% of the flips fell between 40% heads and 60% heads.

Now, suppose we change the size of the class, the sample size. Instead of a class of 10 students, we have a class of 100 students all flipping a coin. Then, we add up all the heads in the class of 100 students. Using the computer to flip the coins for 100,000 classes each of class size 100:

Number of heads	Percentage of heads (%)	Number of classrooms
0–9	0–9	0
10–19	10–19	0
20–29	20–29	0
30–30	30–39	1645
40–49	40–49	4420
50–59	50–59	51,292
60–69	60–69	2858
70–79	70–79	5
80–89	80–89	0
90–100	90–100	0

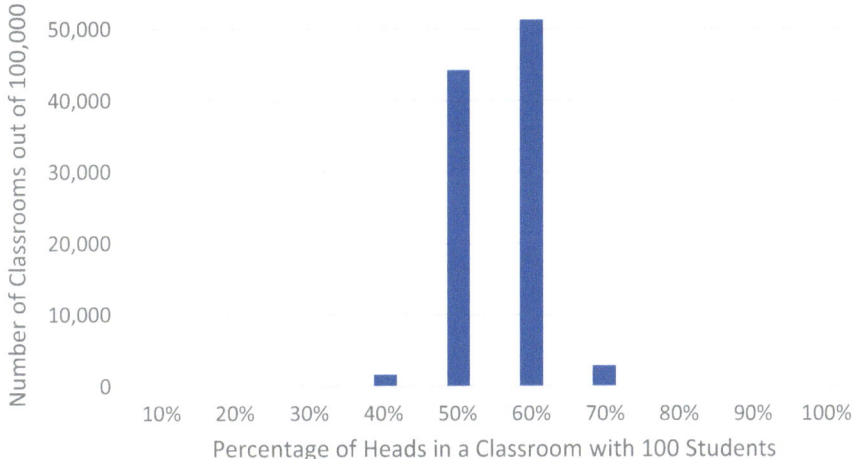

Heads landed exactly at 50 in 8133 trials or 8% of the time. The lowest number of heads recorded is 30 while the highest is 71. And 97% of the trials land between 40% heads and 60% heads. The variance, randomness, uncertainty, luck shrinks with the bigger sample size. No classroom has 100% heads. So, the teachers with the larger classes will see considerably less banners and banquets.

With only 10 students and thus 10 coin flips at a time, getting 10 heads in a row happens. You can get lucky. But with 100 students and 100 coin flips, the odds of getting 100 heads in a row are miniscule. Also, the bulk of trials get closer to the true odds of 50–50, the variance shrinks. With 10 flips, 67% of the trials landed between 40% and 60% heads. But with 100 flips, 97% of the trials landed between 40% and 60% heads. Statisticians call this the law of large numbers which states that as sample size grows, the mean of the sample gets closer to the true mean.

Let's go a step further. Imagine a class that has 1000 students and therefore adds up 1000 coin flips. When we run 100,000 trials with the class of 1000 students the percentages of heads looks as follows:

Number of heads	Percentage of heads (%)	Number of classrooms
0–9	0–9	0
10–19	10–19	0
20–29	20–29	0
30–30	30–39	0
40–49	40–49	48,589
50–59	50–59	51,411
60–69	60–69	0
70–79	70–79	0
80–89	80–89	0
90–100	90–100	0

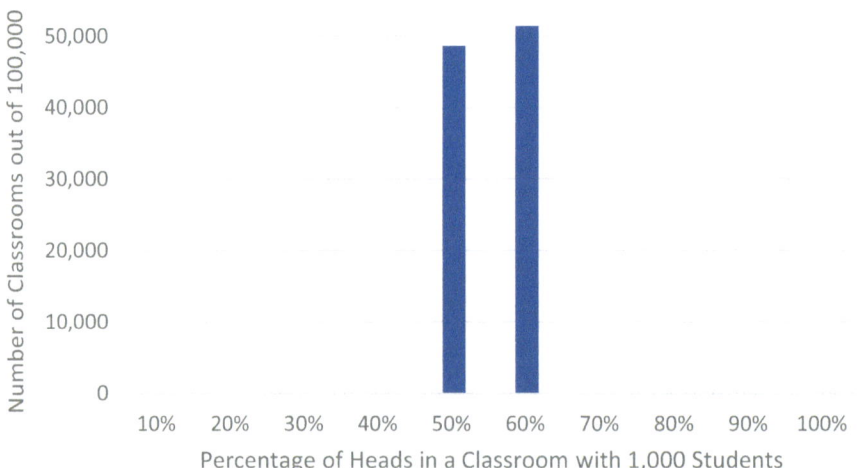

These very large schools see 100% of the classes fall between 40% heads and 60% heads. No banners or banquets for the 1000 student class. They are not close to the 100% heads that we saw in the 10-person class. Give the money to the small schools!

The law of large numbers is intuitive. If we think of height, a 10 person sample with Shaq in the sample is going to overestimate the 5′ 9″ average height. But a 100 person sample with Shaq… not so much. There will be enough short people in the sample to pull down Shaq's length. Shaq's height will not matter at all in a 1000 person sample where many short people will average him out. Or think of batting averages in baseball. Ted Williams is the last person to bat over .400 which he accomplished in 1941 when he had a batting average of .406 with the Boston Red Sox. Nearly every year, a few hitters start the season red hot and carry .400 averages over the first few months of the season. They get lucky. Reporters, pundits, and fans get excited and pontificate on the player's chances of hitting .400. Then, the law of large numbers kicks in and a larger sample size of at bats pulls the averages back down. Their luck runs out.

Small samples will provide strange results. Like the coin flips above, you see 10 heads in a row, but you do not see 100 heads in a row. Or like sampling height with Shaq in the sample. With a sample of 10, Shaq matters… but with a sample of 100, Shaq gets averaged out. A few good students or lucky guesses on the tests will lead to improved test scores… awards and banquets. As long as your school is small enough, a few scores can shift the averages around. But that will not happen in a large school.

To their credit, The Gates Foundation has changed course on supporting small schools. They now advocate "locally driven solutions."[13] The Gates foundation has tremendous resources, but are still capable of being duped by data. They did not fail because their hearts were in the wrong place. They did not fail because their people did not understand the law of large numbers. They did not fail because they lacked effort. They did not fail because they lacked resources. They failed because simple solutions are enticing. We crave them. We get excited, and our emotions cloud our judgment. If all Dukie needed was a smaller classroom, we could fix the problem. We could "fix" education. We could help Dukie. But social problems are not Liam Neeson movies, improving education will take more than 90 minutes.

---

[13] Freedberg, Louis. "Gates Foundation moves to implement new strategy to support 'networks for school improvement,'" Edsource.org, 14 February 2018.

Genius Duped

Daniel Kahneman and Amos Tversky wrote their first paper together in 1971 entitled *Belief in the Law of Small Numbers*.[14] The paper pokes fun at researchers who make too much of interesting results stemming from small samples. Like the teacher whose class gets ten heads, you cannot draw meaningful conclusions from small samples. Kahneman and Tversky followed up this start with many papers that showed how human judgment is often flawed when making decisions under uncertainty. One theme of their work is that nobody is a natural statistician, that statistical understanding is not intuitive. The Nobel Foundation awarded Daniel Kahneman the Nobel Prize in Economics in 2002. Amos Tversky did not receive the award because he died from melanoma in 1996, and the award is not given posthumously.

Kahneman wrote a bestselling book in 2011 highlighting the work that Amos and he did throughout their career called *Thinking, Fast and Slow*.[15] The book is a *NY Times* bestseller, a fascinating read, a darling of the critics, and… the book makes fundamental mistakes interpreting results from small samples. Kahneman highlights work around "priming" in chapter four. He argues that small changes in your environment can change your behavior. For example, one study presented volunteers with stereotypical words related to old people such as "bingo," "wrinkles," and "Florida." Then, the researchers timed how fast the volunteers walked to the door. The study found that people introduced to the words associated with old people walked slower than a control group.[16] The environment can "prime" your behavior.

Kahneman recognized that people may be skeptical of these results, so he wrote in his book:

> *The results are not made up, nor are they statistical flukes. You have no choice but to accept that the major conclusions of these studies are true. More important, you must accept that they are true about you.*

In the years since the book's publication, "priming" effects seem dubious. Replication studies have failed. And worse, some of the original results may be

---

[14] Tversky, Amos, and Daniel Kahneman. "Belief in the law of small numbers." *Psychological bulletin* 76, no. 2 (1971): 105.

[15] Kahneman, Daniel. *Thinking, fast and slow.* Macmillan, 2011.

[16] Bargh, John A., Mark Chen, and Lara Burrows. "Automaticity of social behavior: Direct effects of trait construct and stereotype activation on action." *Journal of personality and social psychology* 71, no. 2 (1996): 230.

fraudulent because researchers fabricated data.[17] Kahneman responded to the controversy with courage and humility. He wrote the following as reported on retractionwatch.com:

> *Clearly, the experimental evidence for the ideas I presented in that chapter was significantly weaker than I believed when I wrote it. This was simply an error: I knew all I needed to know to moderate my enthusiasm for the surprising and elegant findings that I cited, but I did not think it through.*

Nobel prize winners make mistakes with statistics too. As Kahneman taught, we can be seduced by "surprising and elegant" findings. Emotions take over. Statistics are not intuitive. Small samples befuddle. Still, when you get duped by data, make sure you heed Kahneman's most important lesson: act with courage, grace, and humility like he did. Everyone makes mistakes.

CompStat wanted to improve police performance by monitoring crime statistics. The *No Child Left Behind* law thought testing accountability would improve schools. The Gates Foundation wanted smaller schools to be the answer to poor performing schools. Daniel Kahneman was intrigued by how our behavior changes when "primed" by small changes to the environment. But statistics do not provide 90 min of action concluding in happy endings where the good guys win and the bad guys lose. Statistics are nuanced. We want *Taken*; we get *The Wire*. *Taken* ends when Bryan Mills brings his daughter to receive music lessons from a pop star that he protected as part of a security detail. The hero ensures his daughter's happiness. Dukie shoots heroin into his arm in his last scene in *The Wire*. Prez fails to change Dukie's fate. We want simple and elegant; we get complicated and uncertain.

---

[17] Rivers, Andrew M., and Jeff Sherman. "Experimental Design and the Reliability of Priming Effects: Reconsidering the "Train Wreck"." (2018).

9

Nirvana Fallacy

Nirvana Fallacy

Charles Barkley consistently voices his mistrust of sports analytics as a basketball commentator for the NBA on TNT. "People think that I hate analytics. They only do because it's true. I hate analytics," Charles explained on the Dan Patrick Show in 2018.[1] Does Charles have a point?

All statistics are lousy. Statistics provide summaries that mislead. In 1993 when a 38-year-old Moses Malone walked into the Phoenix Suns locker room, Charles declared: "The average age in this room just went up 25 years."[2] Charles's math was slightly off, but his intuition is perfect. The average American family has 1.9 children. The average net worth goes into the billions when Bill Gates walks into the room. Averages often prove foolhardy because averages hide a spread or range of different numbers in a single number, statistics oversimplify.

Basketball field goal percentage equals the number of made shots divided by the total number of shot attempts. Jordan is the all-time leader in field goal percentage with an unreal number of 66.94 percent.[3] Jordan, of course, refers to the three-time All-NBA player DeAndre Jordan of the Brooklyn Nets. DeAndre does not take many shots other than dunks; he has one three-pointer in over 900 games. Michael Jordan comes in number 155 on the list with a pedestrian field goal percentage of 49.69 percent. Basketball ability involves

---

[1] Schwartz, Nick. *Charles Barkley goes off on 'analytical idiots' running NBA teams: 'that's why the league sucks'*, July 19, 2018, USA Today Sports.

[2] Barkley, Charles. *Sir Charles: The Wit and Wisdom of Charles Barkley*, p. 7. Warner Books, Inc., 2009.

[3] As of the end of the 2020 NBA season.

© The Author(s), under exclusive license to Springer Nature Switzerland AG 2024
R. T. Stewart, *Adventures in Statistics*, Copernicus Books,
https://doi.org/10.1007/978-3-031-61284-8_9

more than field goal percentage. Statistics oversimplify. Charles is direct: "analytics is crap."

The advanced analytics folks agree with Charles: field goal percentages are less than perfect. So, they came up with an "advanced" statistic called "effective" field goal percentage or EFG. This stat counts three-pointers 1.5 more than two-pointers because 2 points times 1.5 equals 3 points. EFG gives 50% more credit for making a three-point shot. The advanced EFG stat is better than traditional field goal percentage, but it's still "crap" because it ignores free throws and "and ones." So, stats people invented "true" shooting percentage or TSP which computes the points per attempted shot thus rewarding players for scoring after getting fouled. Michael Jordan moved up the list from 155 in field goal percentage to 95 using TSP. DeAndre Jordan moved down from 1 to 2. The greatest player of all time comes in 95th even with the advanced statistics.

Statistics simplify. This is both a feature and a bug. A single number summarizes a wealth of information, but that summary is a generalization. Field goal percentage along with its siblings EFG and TSP never give a complete picture. Crucial information is left out, like how difficult the shot is being taken or how good your teammates are or what the defense looks like. Michael Jordan garnered a lot of attention from defenses given that he won six championships and five MVP awards. DeAndre Jordan does not garner the same attention given that he averages around 10 points per game on mostly put backs. Charles Barkley has a point.

But Sir Charles may be missing the bigger point. Statistics are guides, not prophecies from the Gods. The 'nirvana fallacy' is the logical error of comparing actual things with idealized alternatives.[4] Politicians often preach of glorified worlds if they are elected, but those utopias never materialize. Perfect solutions do not exist. You cannot eat chocolate cake and ice cream for breakfast, lunch, and dinner… and stay thin. There's a tradeoff. The more sweets you eat, the more likely you are to gain weight. The fantasy of eating whatever you want without gaining weight does not exist. The nirvana fallacy ignores these tradeoffs and pretends that perfect solutions exist. Charles may dream of the perfect statistic that predicts who will win the game, the final score, and who takes home the MVP trophy. But statistics only provide clues, not truths. Statistics do not offer a perfect solution, but statistics do offer a valuable tool to improve decision making.

---

[4] Demsetz, Harold. "Information and efficiency: another viewpoint." *The Journal of Law and Economics* 12, no. 1 (1969): 1–2.

"Analytics Don't Work"

"All these guys who run these organizations who talk about analytics, they have *one* thing in common—they're a bunch of guys who have never played the game, and they never got the girls in high school, and they just want to get in the game," Charles taunted the fans of statistics on TNT in 2015.[5] The lonely stats nerds with poor coordination would likely inform Charles that he mentions *three* things, not *one*.

And the stats guys would agree that field goal percentages, in any form, are limited. Big things—like defense—are ignored. Therefore, the stats Lotharios came up with further complicated statistics. Value over replacement player (VORP) and player efficiency ratings (PER) are two popular stats used to rank players. These advanced analytics measure a whole bunch of stuff to include steals, blocked shots, turnovers, offensive rebounds, defensive rebounds, and assists. Michael Jordan shoots to the top of the lists when VORP and PER are used. Problem solved… not so fast.

VORP and PER work the same way your car insurance works. The insurance company assigns a bunch of points to your characteristics. So, if you have a drunk driving arrest on your record, you pay more. If you take the train to work, you pay less. VORP is the same. If you grab a bunch of offensive rebounds, your VORP and PER go up. If you make a bunch of turnovers, your VORP and PER go down. But by how much? The "how much" is more subjective than many statisticians wants to admit.

The stats guys did not get the girls in high school, but they aced calculus class. Thus, they use mathematical equations to determine the weighting of these characteristics, the "how much." But even math can be fickle. Statistics can contradict. Simpson's paradox—introduced in Chapter One—is the phenomenon when results flip because information, confounding factors, are missing. Charles says: "Analytics don't work."

Two economists explored data from the 2011 NBA playoffs and found several examples of Simpson's paradox.[6] Game 5 of the Spurs victory over the Grizzlies in the opening round series had the following two-point shooting percentages:

[5] Jankens, Evans. *Charles Barkley Calls Rockets GM Daryl Morey "One Of Those Idiots Who Believes In Analytics."*, February 11, 2015, CBS Broadcasting Inc., 2015.

[6] Ma, Y. Zee, and Andrew M. Ma. "Simpson's paradox and other reversals in basketball: examples from the 2011 NBA playoffs." *International Journal of Sports Science and Engineering* 5, no. 3 (2011): 145–154.

Two-point shooting			
Team	Makes	Attemps	Shooting percentage (%)
Grizzlies	38	77	49.4
Spurs	32	63	50.8

The Spurs shot better than the Grizzlies from two-point range. The spurs also shot better than the Grizzlies from three-point range:

Three-point shooting			
Team	Makes	Attemps	Shooting percentage (%)
Grizzlies	3	10	30.0
Spurs	7	22	31.8

Clearly, the Spurs had a better night than the Grizzlies, thus, earning the victory. But what happens when we look at the overall percentages with both two-point and three-point shots:

Combined shooting			
Team	Makes	Attemps	Shooting percentage (%)
Grizzlies	41	87	47.1
Spurs	39	85	45.9

Clearly, the Grizzlies had a better night than the Spurs. What? The hard *facts* that statistical analysis provide… just flipped around. You draw opposite conclusions based on what stats you look at. The stats nerds call this Simpson's paradox. You can make the Grizzlies look good or the Spurs look good depending on which stats you choose. Which one should you use for your VORP or PER calculation? The choice makes a difference, but the answer is ambiguous. Statistics are more subjective than they appear.

An aphorism overused by statisticians says: "Yes, it's easy to lie with statistics, but it's even easier to lie without them." Statistics cannot provide perfect solutions, but the nirvana fallacy teaches that there are no perfect solutions. Statistics help us get closer to the "truth"… provide us with a clearer picture, tell a better story. Statistics provide clues. And without those clues, your case should be thrown out. But do not make statistics your only tool. You'll need many clues to win the case. Charles's argument that "analytics don't work" should be modified to "analytics don't work alone." Always use many methods, many models of how the world works to make your case, to tell the story. "Truth" is elusive.

"The NBA Is About Talent"

The three-point shot won over the analytics crowd and changed the NBA in the 2000s. Throughout the 1980s during the Larry Bird versus Magic Johnson drama, NBA shooters made 23,871 three-pointers. In the 2018–2019 single season, NBA shooters made 27,955 three-pointers. Midrange jumpers made Michael Jordan the GOAT, but the midrange jumper will get you fired in 2022. Charles likes his era: "Listen, you can't win the game shooting threes unless you've got Steph, Klay, and Kevin Durant. Those guys can be analytical. That's the only reason it works."[7] Steph Curry is the greatest shooter the league has ever seen. You can look at the stats; you can watch the highlights; you can count the trophies. Steph falls outside the norm. Charles is right when he tells the analytics people that they are making a mistake when they try to fight fire with fire. You're not going to beat Stephen Curry in a three-point contest.

Basketball analytics—all analytics—suffer from a built-in assumption that historical patterns are a dependable guide to the future. Michael Lewis of *Moneyball* fame wrote an article about basketball analytics highlighting Shane Battier's no-stats genius where he compares basketball analytics to card counting in blackjack.[8] You figure out the odds by keeping track of the cards played, and make large bets when the cards fall in your favor. Unfortunately, historical patterns are not always a dependable guide to the future in basketball… or blackjack.

Ed Thorp invented card counting when he was a young Massachusetts Institute of Technology (MIT) professor cavorting with New York mobsters interested in beating the casinos.[9] His strategy is brilliant and he published an academic article, then, a bestselling book: *Beat the Dealer*. Unsurprisingly, the casinos did not sit back and allow a flood of aspiring card counters in their casinos to drive them bankrupt. The casinos changed the rules. Now, when you count the cards in a casino effectively, they see you changing your bet size with the eye in the sky and politely guide you to a table with an automatic shuffler which makes card counting impossible. Ed Thorp's strategy thwarted. Historical patterns made meaningless; the rules changed.

Basketball teams react to effective strategies just like casinos. They change the rules. The Jordan rules pursued by the Pistons and Don Nelson's

---

[7] Schwartz, Nick. *Charles Barkley goes off on 'analytical idiots' running NBA teams: 'that's why the league sucks'*, July 19, 2018, USA Today Sports.

[8] Lewis, Michael. "The no-stats all-star." *The New York Times* 13 (2009): 1–8.

[9] Thorp, Edward O. *A man for all markets: From Las Vegas to Wall Street, how I beat the dealer and the market.* Random House Incorporated, 2017.

'hack-a-Shaq' defense offer extreme examples. Therefore, the historical patterns found in the analytics will not offer great insight when defenses react. *Beat the Dealer* does not work in the 2024 casinos. The rules change faster than the data.

Numbers can explain only so much. They say nothing about heart and character; they say nothing about unique talent. Unique talent is needed to win championships. Steph won championships by hitting three-pointers like no player in history. And Lebron James won championships with a linebacker's body combined with track star speed and innate passing abilities that nobody had ever seen before. And Tim Duncan won championships with fundamentals and leadership that was incomparable to the past. And Michael Jordan won championships with unbelievable athleticism and drive that has never been duplicated before or since. And Kareem Abdul Jabbar won championships with an indefensible hook shot that has never been copied. And the next guy will win championships with some unique, new talent that we've never seen. Discovering that new talent which wins championships is not conducive to analytical analysis. Analytics explains what has happened in the past. Analytics explain what happens on average. New, unpredictable talent will drive what happens in the future. Players like Zion Williamson, Luka Doncic, Jayson Tatum, and Victor Wembanyama intrigue because they are unique, they have a chance to be that next unique talent. Barkley's most valuable insight is that "the NBA is about talent."

"He's One of Those Idiots Who Believes in Analytics"

"I'm not worried about Daryl Morey, he's one of those idiots who believes in analytics," said Charles Barkley on TNT after a Houston Rockets game in 2015. "First of all, I've always believed analytics was crap. You know I never mention the Rockets as legitimate contenders because they're not. And listen, I wouldn't know Daryl Morey if he walked into this room right now." Morey was the Rocket's general manager at the time, and he fired off the following tweet which riled up Charles: "Best part of being at a TNT game live is it is easy to avoid Charles spewing misinformed vitriol disguised as entertainment."[10] A good old-fashioned feud that probably upped the ratings on TNT.

---

[10] Deitsch, Richard. *Charles Barkley explains feud with Rockets general manager Daryl Morey,* February 13, 2015. Sports Illustrated, 2015.

Daryl Morey earned his MBA from MIT after earning an undergraduate degree in computer science from Northwestern. He started his career as a statistical sports consultant, and rode the *Moneyball* wave to land the general manager job with the Houston Rockets. Michael Lewis highlights Morey's success in his book *The Undoing Project*. Morey took basketball statistics mainstream. But sports statistics geeks know Morey for being founder of the MIT Sloan Sports Conference, also known as "dorkapalooza."

The Sloan conference attracts Nobel Prize winners, best-selling authors, and hall-of-fame athletes to speak about sports and statistics. It's WrestleMania for sports geeks. And in 2014, the main event was long-time champion with multiple best-sellers Malcolm Gladwell versus the young upstart with one successful book to his credit, David Epstein. This feud started when Epstein criticized Gladwell's book *Outliers* for over-hyping the 10,000 hour rule.

The 10,000 hour rule argues that 10,000 hours of deliberate practice leads to greatness. "The idea that excellence at performing a complex task requires a critical minimum level of practice surfaces again and again in studies of expertise. In fact, researchers have settled on what they believe is the magic number for true expertise: 10,000 hours," writes Gladwell.[11] He uses The Beatles, Mozart, Bobby Fischer, and Bill Gates to defend the rule. Epstein suplexes this notion in his book *The Sports Gene*. He argues that there are no "magic" numbers; the 10,000 hours is an average; and you must use a range around the number. "Scientists must go beyond saying that practice matters and attempt the difficult task of determining exactly *how much* practice matters. By the strictest 10,000-hours thinking accumulated practice should explain most or all of the variance in skill. But that never, ever happens," writes Epstein.[12] Yes, practice matters, but genetics matter too.

Like Vince McMahon, the Sloan conference set up steel cages and invited Gladwell and Epstein to fight it out. The match was billed as "10,000 Hours vs. The Sports Gene" on March 10, 2014. The anticipation was palpable. But the battle disappointed. No chairs were thrown. No yelling or screaming. No blood. Not even any mean tweets. Two adults had a civilized conversation, acknowledged each other's points of view … even changed their minds. It was horrible, only a few thousand bothered watching the replays on YouTube. The audience wanted Hulk Hogan slamming Andre the Giant in a throwback to 1987's WrestleMania III. But they got a civilized conversation between mature adults. In the end, both Gladwell and Epstein agreed that it's both nature *and* nurture in the most cliched ending to a steel cage match ever. Worse, Gladwell and Epstein both got it wrong.

---

[11] Gladwell, Malcolm. *Outliers: The story of success*. Little, Brown, 2008.

[12] Epstein, David J. *The sports gene: Inside the science of extraordinary athletic performance*. Penguin, 2014.

The Hidden Half

Dogs have a better sense of smell relative to humans because dogs have more "sensory receptor sites" inside their noses. "Human noses have about six million of these sensory receptor sites; sheepdog noses, over two hundred million; beagle noses, over three hundred million. Dogs have more genes committed to coding olfactory cells, more cells, and more *kinds* of cells, able to detect more kinds of smells. The difference is exponential," writes Alexandra Horowitz in her book *Inside of a Dog*. "We might notice if our coffee's been sweetened with a teaspoon of sugar; a dog can detect a teaspoon of sugar diluted in a million gallons of water: two Olympic-sized pools full."[13] Humans cannot appreciate nor fully understand how dogs smell because the biology is so different. Dogs smell time. If a prisoner runs outside the prison walls, a tracker dog can smell the difference between the first, fifth, and fiftieth footprint simply from the odors coming from the footprints. That's why running in circles will not fool a tracker dog. You smell a rose. Dogs smell every petal on a rose distinctively; dogs can smell whether a human has touched each individual petal on a rose. Dogs can smell if a beetle has landed on a rose petal an hour ago or 3 days ago. Human smell is limited, so much is hidden from us.

Gladwell and Epstein fail to acknowledge what we do not know. What we cannot smell. We smell the rose and believe we understand, but we cannot smell what the dog smells. We cannot smell nor understand what Michael Blastland calls "the hidden half." Gladwell exaggerates the importance of practice, glorifies 10,000 hours, and argues that poverty and attitudes and culture drive success. He's probably right. Epstein points out that heredities and "the sports gene" drive success as much or more than practice. He's probably right too. Nurture and nature both matter, agreed by all. But Gladwell and Epstein fail to acknowledge what we do not know. Not what the dog saw, but what the dog smelled. They do not acknowledge what we miss, how much remains hidden. We can cherry pick stories that fit the narrative, but what is missing? What is the dog smelling that we're incapable of smelling? There is more to the story than nature and nurture and it's hidden from us. We do not know.

A woman's left breast has the exact same genetic makeup as her right breast. Similarly, a woman's right and left breast have been subject to the exact same amount of nutrition, exercise, and sleep. The nurture and the nature are the same for the right breast as they are for the left breast. Yet the right and left breast get cancer at different rates. "Excluding women with BRCA1 and

---

[13] Horowitz, Alexandra. *Inside of a dog: What dogs see, smell, and know.* Simon and Schuster, 2010.

BRCA2 genes, the actual cancer in the other breast is not very much greater than the chance of breast cancer in a woman who has never had it before," writes Blastland.[14] If you have cancer in your left breast, your chance of getting cancer in your right breast remains about the same as the rate for women overall. We do not know why one breast gets cancer while the other does not. Blastland explains, "What goes for breasts also applies to testicles, kidneys, and anything else people have two of, where we also see differences despite a shared history of everything." Cancer may come down to a single cell in a breast. Differences are so subtlety varied, that we as humans cannot determine what those differences could be. We do not know why one breast gets cancer while the other does not. This is the hidden half... we can explain some things, but there is always a *hidden half* that we cannot explain. We know that practice matters and we know that your genes matter... and we know that something else matters but we do not know what or how or why. A smell is there, but we cannot pick it up.

The nirvana fallacy mistakes fantasy for reality, imagination for truth, an idealized solution for a practical solution. Humans often fool ourselves into believing that we understand more than we do. Friedrich Hayek calls it the "pretense of knowledge." Daniel J. Boorstin calls it the "illusion of knowledge." John Anderson King and Mervyn A. Kay call things we can't know "radical uncertainty."[15] Charles Barkley calls it "crap." "What cruel mistakes are sometimes made by benevolent men and women in matters of business about which they can know nothing and think they know a great deal," taught Florence Nightingale over 100 years ago.[16] All truths contain a hidden half that we do not understand, and never will. Unknown. Unknowable.

No theory is ever complete, no statistic foolproof. Nirvana cannot exist. We cannot use statistics to fully understand basketball, much less more complicated issues like economics, healthcare, and love. Statistics are not panaceas in any setting. Statistical methods cannot rid data—or life—of uncertainty. Yet statistics used wisely offer invaluable clues. Decision makers must say we cannot know this, so how do we use the statistics and the facts we do know to improve this decision. We must be humble; we cannot smell like dogs. But we also must understand the power that statistics hold. Statistics—done well—can change the world as great statisticians have proven.

---

[14] Blastland, Michael. *The hidden half: How the world conceals its secrets.* Atlantic Books, 2019.

[15] Kay, John Anderson, and Mervyn A. King. *Radical uncertainty*. Decision-making beyond the numbers: Bridge Street Press, 2020.

[16] Nightingale, Florence. *Notes on nursing what it is, and what it is not.* D. Appleton and Company, 1898.

.

10

Statistics Hall of Fame

Inaugural Class

"To understand God's thoughts, we must study statistics, for these are the measures of his purpose," wrote Florence Nightingale more than 150 years ago. Therefore, divine providence demands a Statistics Hall of Fame. After Fermat and Pascal ventured into the world of what is most likely to happen in the seventeenth century, we used statistics to cure disease, build skyscrapers, fly through the sky, and explore space. Statistics have saved lives, launched industries, and enlightened our lives. Statistics have changed the world. Therefore, we must honor those statisticians whose work led the way.

The Statistics Hall of Fame will preserve history, honor excellence within the field, and connect generations. Statistics is making sense of numbers and nobody has done that better than the members of the inaugural class of the Statistics Hall of Fame. The voting committee considered the person's contributions to the practical application of statistics, and the contributor's integrity, character, and ability. With these criteria, the committee chose five statisticians in the inaugural class: Florence Nightingale, Ronald A. Fisher, Edward O. Thorp, Hans Rosling, and Bill James.

Florence Nightingale (1820–1910)

Florence Nightingale was born into a wealthy family in London in the early nineteenth century where she was expected to look nice, attend church, marry a fellow aristocrat, and make babies. Florence had other ideas. At the time,

© The Author(s), under exclusive license to Springer Nature Switzerland AG 2024
R. T. Stewart, *Adventures in Statistics*, Copernicus Books   ,
https://doi.org/10.1007/978-3-031-61284-8_10

women did not attend university nor pursue careers, but Florence's father educated his curious daughter. She learned Italian, Latin, Greek, philosophy, and history as per the norm, but—at his daughters urging—he also taught her writing and mathematics. Highborn women of the time did not study writing and math. "We are a breed of ducks who have hatched a wild swan," cried Florence's horrified mother.[1]

The wild swan did not stop with mathematics. Florence believed that God had predestined her for the service of nursing. But nursing was not a noble profession at the time. Rather, nursing in Victorian-era England was a mess. Doctor Keith Horsley explains:

> At that time, nursing was seen as the very lowest of vocations. Most nurses were alcoholics; they were permitted and expected to drink alcohol while they worked. Many of them were also prostitutes. It was customary for young women of low social class to look to a life of prostitution and nursing, with the consumption of large amounts of alcohol to make the other two occupations a little more bearable. The average nurse when Florence was young would think nothing of combining her three occupational interests in a single night; she would sit watching over her patient, sipping her gin, and if her patient was well enough and had the money, extra services could be provided.[2]

Florence set about revolutionizing nursing. Her career took off when *The London Times* wrote repeated stories about the wounded and sick British soldiers receiving horrendous care during the senseless Crimean War in Turkey. When *The Times* explained that French hospitals were far superior to the British hospitals, the government acted by recruiting Florence. She took 38 nurses to Istanbul where she took over Scutari Hospital which was infested with fleas, lice, rats, mice, and filth. Wounded and sick soldiers could be found lying in their own excrement. Florence went to work.

Florence and her nurses immediately began cleaning the hospital wards, the kitchens, and the soldiers. She finagled money from donations to The Times, philanthropists, and her own private funds to buy soap, towels, food, sheets, flatware, scissors, and anything else necessary to improve conditions. She dressed wounds, fed soldiers, changed bandages, comforted soldiers, and helped them write letters home. The soldiers worshiped her. She earned the moniker "the lady with the lamp" meaning that she came at night to comfort in full light, not for "extra services." The military brass resented that her authority fell outside military lines and they hated that she was a woman

---

[1] Victorians, Eminent. "by Lytton Strachey." *London: Chatto & Windus* (1918).

[2] Horsley, Keith. "Florence Nightingale." *Journal of Military and Veterans Health* 18, no. 4 (2010): 4.

directing a military hospital like a general. Florence did not give a damn; she was on a mission from God.

But improving sanitation was only part of Florence's work. Florence was a devout statistician who kept track of everything. Before her arrival at Scutari Hospital, the number of deaths was not recorded accurately with separate lists reporting contradictory information. Florence went to work. She started collecting and organizing data not just on the number and causes of death, but also on overcrowding, ventilation, drainage, and cleanliness. Furthermore, she made sense of the data, learning that diseases—often times preventable diseases—were more deadly than battle. She used her data to show that her sanitary reforms were making a difference. She showed that mortality in Scutari Hospital dropped from 42.7% to 2.2% following her reforms.[3] The Times reported Florence's accomplishments, and she became a celebrity across the globe. Florence Nightingale was arguably the most famous woman in the world at the time—excluding Queen Victoria. But soon after her return from war, she would visit the Queen.

Florence returned from the Crimean War determined to use what she learned to improve hospitals and public health. She found preventable deaths inexcusable. "I stand at the altar of the murdered men, and while I live, I fight their cause," she wrote in her memoirs.[4] The weapon Florence used in this fight was statistics. At the time, using statistics for social problems was scandalous and many contemporaries questioned its efficacy. The Charles Barkley of the time was another Charles—Charles Dickens. He wrote *Hard Times* satirizing people who believed in dehumanizing statistics and called statisticians "the representatives of the wickedest and most enormous vice of this time."[5] Florence did not give a damn. She had to convince Queen Victoria, and in this endeavor, she excelled.

Florence knew that she had to make statistics accessible to everyone so that statistics could be reported in newspapers, understood by politicians, and followed by members of royalty. She would ask "Would the Queen understand this point?" She privately referred to this criterion as the "Queen Victoria test."[6] This led to her being the first person to use pie charts in health sciences. She used colored graphics, and invented polar-area charts where the statistic being represented is proportional to the area of a wedge in a circular diagram.

---

[3] Cohen, I. Bernard. "Florence Nightingale." *Scientific American* 250, no. 3 (1984): 128–137.

[4] Kopf, Edwin W. "Florence Nightingale as statistician." *Quarterly publications of the American Statistical Association* 15, no. 116 (1916): 388–404.

[5] Bayley, Mel. "Hard times and statistics." *BSHM Bulletin: Journal of the British Society for the History of Mathematics,* 22, no. 2 (2007): 92–103.

[6] Horsley, Keith. "Florence Nightingale." *Journal of Military and Veterans Health* 18, no. 4 (2010): 4.

She convinced Queen Victoria to establish a royal commission to explore the problems with hospitals during the Crimean War. Then, she worked with other statisticians to implore the army to incorporate changes. She won. Florence drove improvements in ventilation, heating, sewage disposal, water supply, and kitchens. A sanitary code for the army was established and the procedures for gathering medical statistics were standardized. She spent the rest of her life promoting improved health and safety. Only God knows how many lives she saved.

Florence Nightingale is renowned by statisticians for innovation in graphical representations of data. She wrote over 200 pamphlets, papers, books, and articles including the still used *Notes on Nursing*.[7] She is the first woman to fellowship in the Royal Statistical Society, and she was elected an honorary member of the American Statistical Association.

Ronald A. Fisher (1890–1962)

R. A. Fisher revealed his mathematical brilliance at an early age. He attended lectures from leading mathematical professors at the age of seven.[8] His parents and his teachers understood that he had special talents, but he also suffered from severe myopia so his parents forbade him from working under electric light. Thus, his teachers often instructed him without paper or pencils. So, Fisher solved problems in his head, and developed a unique geometrical feel for mathematical problems. For reasons that nobody understands, Ronald's shortsightedness enabled him to "see" problems from a different viewpoint, to learn mathematics slightly askew of his classmates and his professors. The results were stunning. His teachers and professors considered him a genius, and his contributions to the field of statistics are extraordinary. He is referred to as the "father of modern statistics."

Fisher's most successful book is *The Design of Experiments* which was released in 1935 and begins with a story about a lady tasting tea. "A lady declares that by tasting a cup of tea made with milk she can discriminate whether the milk or tea infusion was first added to the cup," writes Fisher.[9] This "experiment" began when Muriel Bristol refused a cup of tea that Fisher prepared by pouring the milk in first. Fisher felt that Muriel's behavior was

---

[7] Nightingale, Florence. *Notes on nursing for the labouring classes*. Harrison, 1861.

[8] Yates, Frank and Mather Kenneth. 1 November 1963, Ronald Aylmer Fisher 1890–1962, Biographical Memoirs of Fellows of the Royal Society, Volume 9, pp. 91–129.

[9] Fisher, Ronald A. "The design of experiments." (1935).

petulant because he believed thermodynamics taught that mixing A and B is no different from mixing B and A. He used the "it's science" argument to determine that Muriel could not taste the difference. Muriel insisted that she could taste the difference and would not drink tea with milk poured in first. To resolve the dispute an experiment was designed where Muriel would randomly be given 8 cups of tea: four with milk poured first and four with tea poured first. Fisher prepared the tea, and randomly presented the cups to Muriel. She got 8 out of 8 cups correct. Fisher was stunned, and intrigued.

His first thought was that she got lucky. If Muriel picks heads and the coin comes up heads, she's not a soothsayer. She got lucky. But suppose she guesses correctly for 5 coin flips in a row… or 10 coin flips in a row… or 100 coin flips in a row. At what point should you believe that Muriel is a soothsayer? How many cups of tea must she guess correctly before we believe her? Fisher developed this reasoning, did the math, and invented concepts around the null hypothesis and statistical significance that are foundational to modern statistics. All because Muriel complained about her tea. But that's not the end of the story. Scientists subsequently worked on the tea and milk problem. It turns out that milk contains proteins that are hydrophobic or water hating, and when water hating proteins encounter water, they curl up into little balls to minimize contact. This process causes "scalding" which results in a change of flavor; the proteins acquire a burnt caramel flavor. If you pour tea into milk, the scalding cannot occur because the little balls are not formed. But when the milk comes second, the scalding occurs, and the caramel flavor is revealed. Muriel was correct.

Fisher did not earn the moniker "father of modern statistics" by just creating the concepts of statistical significance and the null hypothesis although that probably would have been enough. His contributions to statistics read like a Stats 101 syllabus. He demarcated sample means versus population means, defined randomness, created the statistical method of analysis of variance (ANOVA), introduced the concept of maximum likelihood, developed the F-distribution, and much, much more. Nobody has contributed more to the statistics canon then R. A. Fisher.

But Ronald was imperfect, his temper notorious. Professor J.F. Crow explains:

> *Being frustrated, he erupted. At one moment in fury against a laboratory assistant, he crushed to death the mouse in his hand, then, realizing, muttered "See what you have made me do," cast the mouse from the open window and left the room; but a*

moment later, re-opening the door and seeing the girl facing him in an attitude of rude gesticulation, he grinned, friendly again, and slipped away.[10]

And Fisher was deeply involved in eugenics, the study of improving the human race through selective breeding. He headed the Department of Eugenics at University College London in the 1930s. In 2021, UCL issued a formal apology recognizing their role in promoting "the spurious idea that varieties of human life could be assigned different value" and that this backing "provided justification for some of the most appalling crimes in human history: genocide, forced euthanasia, colonialism, and other forms of mass murder and oppression based on racial and ableist hierarchy."[11] As head of the UCL eugenics department, RA Fisher was front and center in the eugenics movement, and his oversized intellect and mastery of statistics failed to save him from supporting these spurious and dangerous ideas. Fisher's role in eugenics marks a nadir for statistics. Worse, Fisher was not alone. Other intellectuals who made important contributions to statistics—Francis Galton, Karl Pearson, and John Maynard Keynes—accepted eugenic ideas too. Why didn't the number save these giants in the field? Why didn't science point the way? Why didn't statistics work?

Science does not provide black and white answers wrapped up in unfettered conclusions. Numbers too often provide the illusion of certainty, the myth that we can separate fact from fiction, truth from false, spurious from significant, right from wrong. But the world is much too complicated. Scientific knowledge is not irrefutable facts, but tentative hypotheses subject to constant revision. Yes, science is evidence based, but the evidence never comes in the form of Perry Mason waltzing into the debate with a smoking gun. There is no algorithm affirming whether your hypothesis is true or false. Rather you only see the data, data that is sparse and unclear. Science is filled with uncertainty, question marks, randomness, and unknowns. Science is dynamic, never certain. And science is entirely human, complete with the foibles and prejudices that make us human. Scientists struggle to make sense of foggy evidence with unknown answers, and they come to their analysis with preconceived beliefs, feelings, and prejudices. Harvard professor Stephen Jay Gould explains that science can never be impartial: "I criticize the myth that science itself is an objective exercise… I believe that science must be understood as a social phenomenon, a gutsy, human enterprise not the work

---

[10] "R.A. Fisher—50 Years Beyond." *Significance—The Royal Statistical Society*, December 2012.
[11] UCL Apology

of robots programmed to collect pure information."[12] Science is a human endeavor replete with human failures even for people as intellectually gifted as Ronald Fisher.

Fisher defended smoking and promoted eugenics, and we look at this with contempt. But we can learn from Fisher's failures. Science, including statistics, should always be laced with a healthy dose of humility because science can never be separated from feelings, class, and culture.

Edward O. Thorp (1932–)

Edward O. Thorp was born in the midst of the great depression. His father was a war hero who had earned a Bronze Star, Silver Star, and two Purple Hearts in World War One. But during Ed's early childhood, Ed's father was struggling to support a wife and two children on a $25-a-week salary. And his oldest son had yet to speak a word at nearly 3 years old. Ed's parents took their reticent boy to the doctor who told them that the boy would talk when he was ready. Around his third birthday, Ed was ready. He began to speak… in complete sentences. He learned to read soon after and was reading college level books while still in single digits. His curious mind led him to chemistry, physics, and mathematics. He earned his PhD from UCLA and landed a teaching job at MIT. His parents were proud.

Ed is a legend in the gambling community because of his adventures with both blackjack and roulette discussed in chapters two and five. His gambling winnings and proceeds from his best-selling book on card counting—*Beat the Dealer*—gave Ed a small bankroll that led to his interest in investing.[13] He devoured every book on the topic and used his gambling skills to invent a mathematical strategy to beat the market. He was the world's first quant, and he revolutionized Wall Street.

Ed founded Princeton Newport News, the first quantitative hedge fund. The fund averaged 19.1% annual returns for two decades which translates to a $100 thousand investment growing to more than $3 million in 20 years. The fund had no down years, so was truly market neutral meaning that the fund made money regardless of what happened in the market. Thorpe did this by taking advantage of mispriced derivative contracts, primarily warrants. He used a precursor to the Black-Scholes formula that allowed him to take advantage of pricing differences between stocks and the options that derive their

---

[12] Gould, Stephen Jay, and Steven James Gold. *The mismeasure of man*. WW Norton & company, 1996.

[13] Thorp, Edward O. *Beat the dealer: A winning strategy for the game of twenty-one*. Vol. 310. Vintage, 1966.

value from those stocks. He pioneered strategies that modern hedge funds use to this day including option, convertible, index, and statistical arbitrage. Ed explains his strategies in another best-selling book called *Beat the Market*.[14] Ed retells his life story in yet another best-selling book called *A Man for All Markets* which reads like a James Bond movie.[15] He has seen and done it all: gambling guru, financial innovator, and author extraordinaire.

Hans Rosling (1948–2017)

Hans Rosling was born July 27, 1948, in Uppsala, Sweden. Hans explains his family's fortunes in his book *Factfullness*:

> *I was four years old when I saw my mother load a washing machine for the first time. It was a great day for my mother; she and my father had been saving money for years to be able to buy that machine. Grandma, who had been invited to the inauguration ceremony for the new washing machine, was even more excited. She had been heating water with firewood and hand-washing laundry her whole life. Now she was going to watch electricity do that work. She was so excited that she sat on a chair in front of the machine for the entire washing cycle, mesmerized. To her, the machine was a miracle.*

> *It was a miracle for my mother and me too. It was a magic machine. Because that very day my mother said to me, "Now, Hans, we have loaded laundry. The machine will do the work. So now we can go to the library." In went the laundry, and out came books. Thank you industrialization, thank you steel mill, thank you power station, thank you chemical-processing industry, for giving us time to read.*[16]

Hans's mother began reading free books to him borrowed from the public library. He became the first in his family to get more than 6 years of education. He got a lot more than 6 years; he became a doctor. He studied medicine and statistics, then, he worked in Africa for years before he became famous… for swallowing swords.

As a child, Hans loved the spectacle of the circus, the impossible became possible. And as a medical student he learned that pushing the chin bone forward allows for straight passage… perfect for swallowing swords. So, Hans

---

[14] Thorp, Edward O. *Beat the Market: A scientific stock market system*. Random House, 1966.

[15] Thorp, Edward O. *A man for all markets: From Las Vegas to wall street, how i beat the dealer and the market*. Random House, 2017.

[16] Rosling, H. with O. Rosling and A. Rosling Ronnlund. 2018. Factfulness: Ten Reasons We're Wrong about the World—and Why Things are Better than You Think."

learned to swallow swords. "I swallow the sword because I want the audience to realize how wrong their intuitions can be. I want them to realize that what I have shown them—both the sword swallowing and the material about the world that came before it—however much it conflicts with their preconceived ideas, however impossible it seems, is true."[17]

Hans used the sword swallowing and the "best stats you've ever seen" to teach us… good news. Journalists don't sell good news because good news unfolds slowly while bad news is often sudden, dramatic, and made for head-lines. If it bleeds, it leads. So, we tend to believe the world is getting worse and the good old days were actually good—what Hans calls our negativity instinct. Don't tell anyone, but the world is getting better. We're inundated with jour-nalism in the internet era, and journalists make money selling drama. Hence, bad news gets larger audiences and conflicts dominate. To sell good news you need someone like Hans—a man with contagious enthusiasm, a man who swallows swords.

Hans asked questions:

> In 1980, roughly 40% of the world's population lived in extreme poverty, with less than $2 per day. What is the share today?[18]

Then he shocks his audience by explaining that the answer is less than 10%. The world is richer and safer than it has ever been. We're making progress with vaccines, life expectancy, education, suicide rates, and women's rights. Han's lectures teach these trends with amazing statistics shown in visual displays that would make Florence Nightingale proud. His stats have received millions of hits on YouTube where his lectures enthrall audiences. His book—*Factfull-ness: Ten Reasons We're Wrong about the World, and Why Things are Better than you Think*—sold out across the globe, taught statistics to a general audience, and influenced laymen and leaders. Hans is the greatest teacher the stats world has ever known.

Bill James (1949–)

Bill James was raised in Mayetta, Kansas—population 341—about an hour and a half drive west of Kansas City, Missouri. His mother died when he was five, so his father raised him while working as a janitor and handyman. Bill

---

[17] Ibid.
[18] Gapminder.org

attended the University of Kansas where he earned a Bachelor of Arts degree in English and economics. He spent a few years in the US Army before settling into a low-paying job as a boiler attendant/security guard at the Stokely-Van Camp's pork and beans cannery in Lawrence, Kansas. The job was perfect for Bill. "I'd spend 5 minutes an hour making sure the furnaces didn't blow up and 55 minutes working on my numbers," explains Bill.[19] His numbers refer to baseball statistics. He brought his baseball encyclopedia and the baseball box scores into his job; and he started asking questions: *At what age do baseball players peak? Does stealing improve your offense? Is baseball 75% pitching?* Then, he looked to the numbers for answers. His results not only changed baseball, but sports across the globe.

Bill paid for a tiny advertisement in *The Sporting News* promoting "*The 1977 Baseball Abstract Contains 18 Statistical Categories That You Can't Find Anywhere Else.*" He asked for $3.50 plus $0.50 postage and handling in exchange for the "abstract" which consisted of 68 pages of statistics and commentary stapled together like an 8th grade report by the kid who waited until the night before to finish the project. Bill sold 75 copies, but was undeterred. His sales steadily increased: 325 in 1978, 600 in 1979, and 750 in 1980. Bill was garnering attention. Like Frosty's old silk hat, there was a little magic in his numbers, stats, and words. Bill's enthusiasm for the subject, curiosity, and snarky honesty intrigued baseball devotees. Plus, he irked baseball elites. His audience grew. In 1981, Sports Illustrated did a piece on Bill's numbers refuting baseball norms, and Bill's little project grew legs. Sports analytics was born.

Over the years, Bill's abstract grew from stapled pages to an actual bound book, and Bill started making money selling his statistics. He developed countless statistical innovations that are commonplace for baseball teams today including: runs created, range factors, defensive efficiency ratings, win shares, Pythagorean winning percentages, game scores, and similarity scores. But Bill's influence skyrocketed when Michael Lewis released his book *Moneyball: The Art of Winning an Unfair Game.*[20] Baseball went bananas for statistics, and the rest of the sports world soon followed. Forget about Michael Jordan, everyone wanted to be like Bill James. He made statistical analysis of sports performance the norm; he changed the world.

"Our capacity to misunderstand the world is almost without limit," Bill explained to a room full of well-educated statisticians.[21] Bill's genius derives from this viewpoint. "The reality is that we're not capable of understanding

---

[19] Okrent, Daniel. "He does it by the number." *Sports Illustrated,* 25 May 1981.

[20] Lewis, Michael. *Moneyball: The art of winning an unfair game.* WW Norton & Company, 2004.

[21] *Battling Expertise with the Power of Ignorance,* Bill James, 14 April 2010, billjamesonline/article1372.

the world, because the world is vastly more complicated than the human mind." Bill gets how the dog smells. To combat our ignorance, Bill asks seemingly simple questions, and devises ways of measuring anything that may answer the question. That is what statistics is all about. Bill goes on to remind his audience of PhDs, "I have no credentials whatsoever as a mathematician or a statistician." You do now Bill; you're a hall of famer.

Statistical Adventures

Statistics was born from inquisitiveness, from wondering, from a desire to understand, to examine, to investigate. Florence Nightingale, Ronald Fisher, Ed Thorp, Hans Rosling, and Bill James don't have a lot in common—different centuries, interests, educations, and upbringings. But they share curiosity as a character trait, an unflinching desire to understand their worlds. That's the most important characteristic behind the success of all the inaugural members of the statistics hall of fame, the most important characteristic for anyone creating or using or thinking about statistics, the most important characteristic for anyone looking to change the world: the yearning to investigate, to understand, to explore. Go.

Epilogue: Top Ten Podcast Episodes About Statistics

Top Ten

Podcasts about statistics cannot compete with Oprah's soulfulness, or horrifying true crime stories, or Joe Rogan's experiences. But the following are the ten best podcast episodes about statistics as of September 2022. You can hear the voices of many of the heroes in this book, in these podcast episodes.

10. *Ed Thorp, The Man Who Beat the Dealer and The Market*
    Masters in Business with Barry Ritholtz; July 13, 2017; 101 Minutes.

The avuncular Barry Ritholtz interviews the legendary statistician Ed Thorp about his life and endless adventures. Ed Thorp was bankrolled by a mafioso to travel to Las Vegas and beat the casinos at blackjack. Then he took on roulette and beat the casinos, again. Then, he created statistical arbitrage in stock markets leading the way for all future hedge funds. Ed's life and accomplishments and adventures are the stuff of movies.

Episode Highlights

Ed describes meeting Warren Buffet through a mutual friend at dinner. He left the dinner and told his wife that the "young" Warren would be the richest person in the world. Soon after, Ed invested in Berkshire Hathaway. Ed ends the interview with: "The last time I was bored, I think I was eleven."

© The Editor(s) (if applicable) and The Author(s), under exclusive license to Springer Nature Switzerland AG 2024
R. T. Stewart, *Adventures in Statistics*, Copernicus Books,
https://doi.org/10.1007/978-3-031-61284-8

9. *Hans Rosling—the Extraordinary Life of a Statistical Guru*
   More or Less: Behind the Stats; February 12, 2017; 26 Minutes.

Tim Harford uses his "More or Less" podcast to explain statistical information found in everyday news. This episode details the life and too early death of the brilliant Hans Rosling. Tim collects stories from Hans's friends and colleagues about Hans's impact. Hans redefined our view of the world, and our use of statistics. Hans taught that statistics can improve humanity.

Episode Highlights

The audio of Hans giving his speeches needs to be heard. Hans makes public health statistics exciting. Passion explodes from his voice: statistics are important; facts are important; you can change the world. Hans demanded that his audience pay attention. "We can no longer have two boxes in our heads: one labeled the western world, the other the developing world." Hans inspires our souls.

8. *How to be Less Terrible at Predicting the Future*
   Freakonomics: Episode 233, January 14, 2016; 47 Minutes.

Author Stephen Dubner interviews Philip Tetlock about his best-selling book, *Superforecasting: the art and science of prediction.*[1] Tetlock studies what makes people skilled at assigning probabilistic forecasts. The experts tend to fail. Rather, people who win forecasting challenges are average. The winners have curious minds and healthy doses of humility, but slight knowledge of the topics they are forecasting. Tetlock dubs these people "superforecasters" and explains what makes them successful. Hard work, curiosity, and skepticism make winners, while intelligence and quantitative skills play a lesser role.

Episode Highlights

What makes the episode superb is that Dubner does not just interview Tetlock. He also talks to the actual superforecasters. Superforecaster Bill Flack explains his method: "I had to dig for background information…I spent a lot of time

---

[1] Tetlock, Philip E., and Dan Gardner. *Superforecasting: The art and science of prediction.* Random House, 2016.

with Google news…I had to educate myself up on the subject." He beat the experts.

7. *What Causes What?*
    Planet Money: Episode #453; October 17, 2018; 19 Minutes.

Jacob Goldstein and Robert Smith are the longtime hosts of the entertaining podcast, *Planet Money*. In this episode, they interview *Naked Statistics* author Charles Wheelan.[2] The episode deliberates on correlation versus causation using amusing stories. Even experienced statisticians will be entertained.

Episode Highlights

Charlie Wheelan explains the horrifying results of a Harvard study that found a link between estrogen therapy and lower rates of heart disease. The study led to thousands of women being prescribed estrogen. A follow up randomized study found "that the women who got estrogen therapy in this controlled experiment actually had higher rates of heart attack, stroke, breast cancer, and other things that were quite damaging … estrogen was actually bad for the health of women." Mistakes interpreting statistics can kill you.

6. *Field of Ignorance*
    Against the Rules; April 12, 2022; 39 Minutes.

Michael Lewis interviews hall-of-famer Bill James about his experiences changing baseball norms and the concepts in Lewis's classic book *Moneyball*. The book became a popular movie starring Brad Pitt which scores 94% on Rotten Tomatoes. Making the topic of statistics into a successful movie gets you on this list.

Episode Highlights

Michael asks Bill why the people running the baseball teams weren't thinking clearly. "Why weren't they thinking, Michael? They weren't thinking because they thought they knew, that's what destroys me and all of us…keeps us from doing the things we could do, thinking that we know things we don't actually

---

[2] Wheelan, Charles. *Naked statistics: Stripping the dread from the data*. WW Norton & Company, 2013.

know," responds Bill. He understands that the dog is picking up smells that humans can't.

5. *John Ioannidis Discusses Why Most...*
     STEM-Talk: Episode 77; November 19, 2018; 85 Minutes.

Stanford professor John Ioannidis's famous paper, "Why Most Published Research Findings are False," forced all statisticians to question the credibility of their work.[3] Dr. Ioannidis uses examples from healthcare to explain how to improve research. One takeaway: practitioners should focus on plausible theories rather than exciting results. John's message should be heard by all who use statistics.

Episode Highlights

Pulling no punches, John questions the "publish or perish" system that may lead to poor healthcare results. "Research is extremely difficult, it's a very noble exercise. It takes a lot of effort, a lot of time. You're open to zillions of potential biases and things that can go wrong," warns John.

4. *Nate Silver on the Supreme Court and the Underrated Stat for Finding Good Food*
     Conversations with Tyler: Episode 7, February 22, 2016; 81 Minutes.

Renowned economist and blogger (marginalrevolution.com) Tyler Cowen asks thought-provoking questions that bring out the best in his guests. Nate Silver does not disappoint. Silver gained fame by predicting 49 of the 50 states in the 2008 presidential election. Then, Silver wrote a best-selling book about predictions, *The Signal and the Noise*.[4] Listening to these two heavyweights jab, shuffle, and punch leaves you demanding a rematch.

Episode Highlights

Tyler challenges Nate throughout the conversation to justify his use of statistical models. Nate's response: "The long way of saying this is that I'm not sure that I'm any better than the average pundit unless I have a model. The

---

[3] Ioannidis, John PA. "Why most published research findings are false." *PLoS medicine* 2, no. 8 (2005): e124.
[4] Silver, Nate. *The signal and the noise: Why so many predictions fail-but some don't*. Penguin, 2012.

disciplining effect of a model, doing your thinking in advance, and setting up rules of evidence is probably quite important."

3. *On Average*
   99% Invisible: Episode 226; August 23, 2016; 19 Minutes.

The average is the most popular statistic of all time. We use averages daily without thought: average height, grade point average (GPA), average income, and batting average. But Todd Rose joins the folks at the well-produced 99% Invisible podcast to shatter people's preconceived notions of "averages." Rose argues that we must reconsider our use of the average. Pulling from his book, *The End of Average,* Rose uses historical anecdotes to drive his message.[5]

Episode Highlights

Adolphe Quetelet gained celebrity in the 1840s by popularizing the use of averages to measure everything from chest circumference to suicide rates. Abraham Lincoln was a fan. Lincoln ordered a massive study of his troops which was used to inform the rationing of food, weapons, and uniforms. Averages won the war.

2. *Think Fast with Daniel Kahneman*
   Hidden Brain; March 12, 2018; 49 Minutes.

Daniel Kahneman visits NPR to discuss his Nobel Prize winning work and friendship with Amos Tversky. Kahneman tells stories about Nazi SS officers, clarifies his views on climate change, and explains his research on biases. He does this with good humor and genuineness. Sadly, Nobel Laureate Daniel Kahneman passed in March 2024, but you can still hear one of the all-time greatest in this wonderful interview.

Episode Highlights

"We found our mistakes very funny, what was fun was finding yourself about to say something *really* stupid," explains Kahneman when describing his relationship with Amos Tversky. These "stupid" thoughts led to the insights that won the Nobel prize. His humor, stories, and knowledge are worth a listen.

---

[5] Rose, Todd. *The end of average: How to succeed in a world that values sameness.* Penguin UK, 2016.

1. *Andrew Gelman on Social Science, Small Samples, and the Garden of Forking Paths*
   Econtalk, March 20, 2017, 68 Minutes.

Russ Roberts is the Godfather of statistical podcasts. Russ' Econtalk.org archives contain a treasure trove of podcasts on statistics running back to 2006. Russ's skepticism and intuitiveness make the conversations captivating. And Don Roberts keeps the often high-level topics assessable to a general audience because of his innate teaching ability. This episode with Columbia professor Andrew Gelman tackles the confusion around statistical significance, p-hacking, and sample size.

Episode Highlights

Professor Gelman explains p-hacking: "Basically, there are many different analyses that you could do of any data set. And you kind of only need *one* that's statistically significant to get it to be publishable. It's a little bit like you have a lottery where there's a 1 in 20 chance of a winning ticket, but you get to keep buying lottery tickets until you win."